# Key Technologies on New Energy Vehicles

Key Technologies on New Energy Vehicles publishes the latest developments in new energy vehicles - quickly, informally and with high quality. The intent is to cover all the main branches of new energy vehicles, both theoretical and applied, including but not limited:

- Control Technology of Driving System
- Hybrid Electric Vehicle Coupling Technology
- Cross Disciplinary design optimization technology
- Single and Group Battery Technology
- Energy Management Technology
- Lightweight Technology
- New Energy Materials and Device
- Internet of Things (IoT)
- Cloud Computing
- 3D Printing
- Virtual Reality Technologies

Within the scopes of the series are monographs, professional books or graduate textbooks, edited volumes, and reference works purposely devoted to support education in related areas at the graduate and post-graduate levels.
To submit a proposal or request further information, please contact:
Dr. Mengchu Huang, Senior Editor, Applied Sciences
Email: mengchu.huang@springer.com
Tel: +86-21-2422 5094

More information about this series at https://link.springer.com/bookseries/16377

Donghai Hu

# Design and Control of Hybrid Brake-by-Wire System for Autonomous Vehicle

Springer

Donghai Hu
School of Automotive and Traffic
Engineering
Jiangsu University
Zhenjiang, Jiangsu, China

ISSN 2662-2920 ISSN 2662-2939 (electronic)
Key Technologies on New Energy Vehicles
ISBN 978-981-16-8945-1 ISBN 978-981-16-8946-8 (eBook)
https://doi.org/10.1007/978-981-16-8946-8



This Springer imprint is published by the registered company Springer Nature Singapore Pte Ltd.
The registered company address is: 152 Beach Road, #21-01/04 Gateway East, Singapore 189721, Singapore

# Preface

The autonomous vehicle is a complex electromechanical integration system. With the development trend of electric and intelligent vehicles, the braking system of it will gradually develop to brake-by-wire in order to achieve a more precise and controllable braking effect. At present, autonomous cars are equipped with hybrid brake-by-wire systems, such as a composite brake system for motor and friction and an integrated brake system for electromagnetic and friction. As a researcher in the field of autonomous vehicles, I have frequently felt that we need a monograph that gives a comprehensive and detailed introduction to the hybrid brake-by-wire system for autonomous vehicles.

This book focuses on the hybrid brake-by-wire system for autonomous vehicles and presents a comprehensive introduction and research on the brake system's modeling, design, and control. This book is composed of six chapters. The first chapter introduces the industry status of hybrid brake-by-wire system, the second chapter introduces the research status of hybrid brake-by-wire system, the third chapter models hybrid brake-by-wire system, and the fourth chapter designs the hybrid brake-by-wire system based on the principles of safety and energy-saving, The fifth chapter puts forward the braking force distribution control method of hybrid brake-by-wire system, and the sixth chapter introduces the mode switching control of hybrid brake-by-wire system.

The book is written for engineers of hybrid brake-by-wire system for autonomous vehicle, teachers and students majoring in automobile and automation. We hope that this selection of materials can provide readers with some guidance in the future design and control of hybrid drive-by-wire systems for autonomous vehicles.

This book is sponsored by the National key research and development program of China "Integrated transportation and intelligent transportation" (2018YFB1600303).

Zhenjiang, Jiangsu, China

Donghai Hu

# Contents

# Chapter 1
# Industry Status of Hybrid Brake-By-Wire System

The system integration of brake by wire and standard hydraulic brake or electro-hydraulic brake has become a trend in developing automobile brake systems in the future. There are two typical automobile integrated brake systems: composite brake system for motor and friction and integrated brake system for electromagnetic and friction. This chapter also describes the structure and working principle of four kinds of brakes.

## 1.1 Motor Brake

As shown in Fig. 1.1 [1], regenerative motor brake (RB) includes drive motor, motor controller, power battery, and battery management system. The drive motor has a four-quadrant working capacity. The x-axis represents the direction of the motor rotating magnetic field, and the y-axis represents the direction of motor load torque. When the driving motor works at the first and third image limits, the driving power is output, the first quadrant is forward driving, and the third quadrant is reverse driving: When the driving motor works at the second and fourth quadrants, the braking power is output. The second quadrant is forward braking, and the fourth quadrant is reverse braking.

When the vehicle is braked, the wheel will drive the drive motor to rotate through the transmission device, and the whole vehicle controller will send a regenerative braking command to the drive motor. At this time, the drive motor will implement motor regenerative braking. On the one hand, the vehicle braking energy will be recovered and stored in the power battery, and on the other hand, the braking torque will be applied to the wheel. As a non-contact braking mode, motor regenerative braking can save energy consumption, improve vehicle driving range, and have no braking dust and noise in the braking process. It is an environment-friendly braking mode. However, the implementation of motor regenerative braking is affected by

 
D. Hu, *Design and Control of Hybrid Brake-by-Wire System for Autonomous Vehicle*, Key Technologies on New Energy Vehicles,
https://doi.org/10.1007/978-981-16-8946-8_1

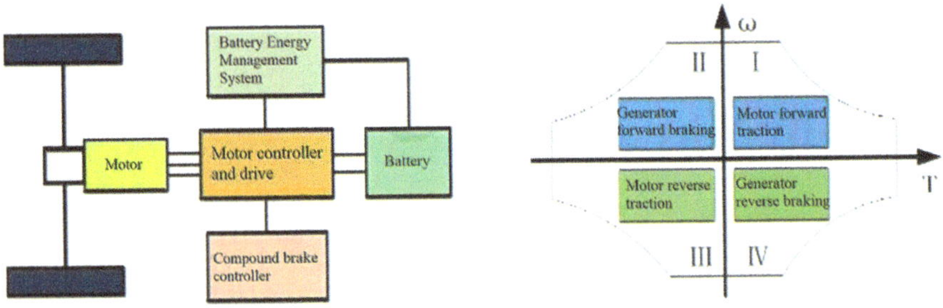


**Fig. 1.1** Structure of motor regeneration brake and four-quadrant operation characteristics of motor

many factors: first, when the state of charge (SOC) of power battery exceeds the upper limit, the single voltage and total voltage of the power battery exceed the upper limit, or the single temperature of power battery exceeds the upper limit, the motor regenerative braking must stop working; Secondly, the vehicle controller needs to obtain the ability of motor regenerative braking to output braking torque according to the driver's braking demand, vehicle speed, the external characteristic curve of driving motor, current gear of transmission, the maximum allowable charging current of power battery and other factors.

## 1.2 Electromagnetic Brake

As a non-contact braking mode, electromagnetic brake (ECB) has the advantages of no friction, no noise, simple control, and rapid response. It mainly includes three parts: electromagnetic brake, electromagnetic brake controller, and battery.

As shown in Fig. 1.2 [2], the electromagnetic brake is composed of the iron core, excitation coil, stator support, and electromagnetic brake disc. The iron core and the excitation coil form an excitation winding, and a plurality of excitation windings install on the stator support. The excitation winding will create an excitation magnetic field if the current is applied to the excitation coil. The excitation magnetic field forms a loop between the electromagnetic brake disc, air gap, and iron core. At this time, countless vortex-induced currents (starting now referred to as "eddy currents") will be generated on the inner surface of the electromagnetic brake disc when it rotates continuously. On the basis of Faraday's law of electromagnetic induction, the excitation magnetic field will produce a braking force on the electromagnetic brake disc with eddy current to prevent it from continuing to rotate. The electromagnetic brake controller receives the electromagnetic brake target braking torque command from the whole vehicle controller and changes the size and form of current (DC or AC) in the excitation coil according to the wheel speed information fed back by the sensor control the output electromagnetic braking torque.

Electromagnetic brake has the advantages of low manufacturing cost, rapid braking response, simple control, no braking noise, and dust pollution. Similar

**Fig. 1.2** Experimental device to test the ECB's performance designed by professor Lee and professor Park

to motor regenerative braking, electromagnetic brake, as a non-contact braking method, can not output ample braking torque and realize vehicle braking and parking; However, compared with motor regenerative braking, the implementation of electromagnetic brake and the output of electromagnetic braking torque are less limited. At the same time, it must also be pointed out that the electromagnetic brake needs to consume a lot of electric energy during operation, which will cause additional fuel consumption for traditional fuel vehicles.

## 1.3 Electro-Hydraulic Brake

At present, the electro-hydraulic brake system has become one of the research hotspots in vehicle active safety and electronics at home and abroad. Petruccelli et al. [3] proposed the control strategy of the front and rear axle braking force distribution of the EHB system and the brake pedal feeling simulation method based on this control strategy under the condition of ordinary braking. At the same time, they proposed the control strategy and self-diagnosis control method under the condition of sensor failure. Finally, they studied the use of an electro-hydraulic brake system to improve the control performance of ABS/VDC. Reuter DF and others also compared and analyzed the brake systems between electronic and traditional hydraulic under the condition of ordinary braking. They explored EHB in different hydraulic design environments and mainly discussed the braking performance analysis of the electro-hydraulic brake system using the standby braking circuit when the energy supply device fails. Hu, K, and others put forward the error diagnosis model of electro-hydraulic brake system to detect the braking state of the system. Based on the model, simulation research was carried out to verify the proposed self-diagnosis control strategy.

Although the electro-hydraulic brake system has been installed and used on some foreign models, such as Mercedes Benz SL and E-class cars, Toyota Prius and Lexus hx400h, at present, China has not fully mastered the EHB design theory and method of the electro-hydraulic brake system, which hinders the industrialization process of the electro-hydraulic brake system in China to a certain extent. The design task of an electro-hydraulic brake system is very complex, and the design of its main components is also affected by many factors.

Figure 1.3 [3–6] shows EHB hydraulic control units from different companies. As shown in Fig. 1.3d, the electro-hydraulic brake system developed by Toyota company

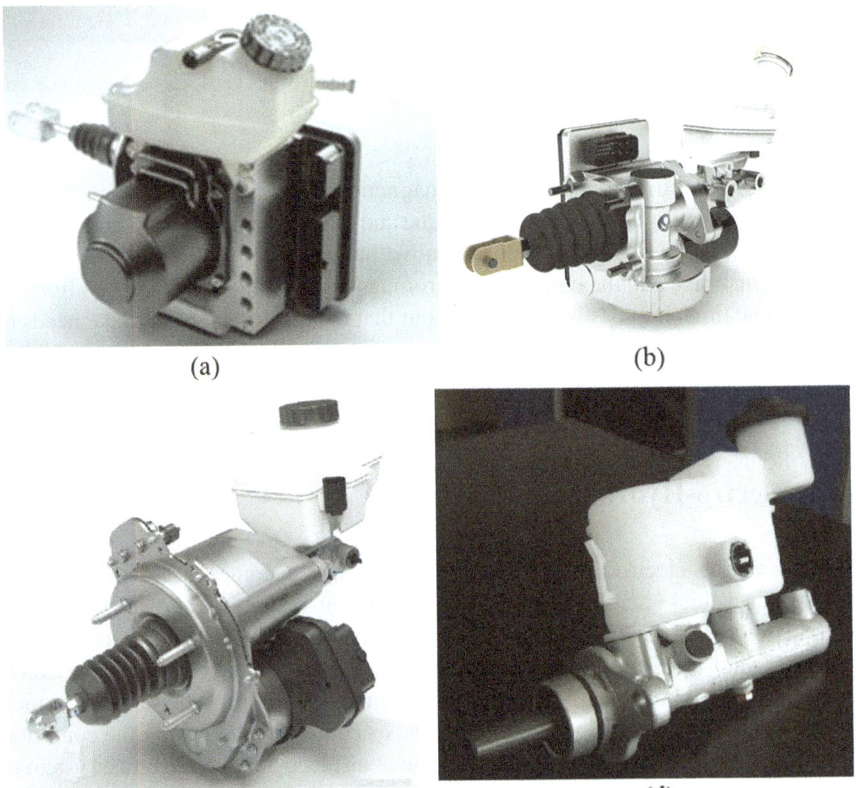

(a) (b) (c) (d)

**Fig. 1.3** Appearance of several EHB hydraulic control units

improves the structure of high-pressure accumulators by adopting compressed metal bellows. When the pressure in the high-pressure accumulator increases due to the small molecular spacing of metal bellows materials, nitrogen infiltration into brake fluid can be effectively solved compared with the scheme of using a rubber diaphragm. As shown in Fig. 1.3a [4], unlike the technical solutions adopted by Bosch and Toyota, Continental integrates an integrated brake master cylinder on the hydraulic regulator.

Compared with traditional hydraulic brake, electro-hydraulic brake has the following advantages: electro-hydraulic brake can significantly shorten the braking response time and reduce the braking distance in case of emergency braking; Electro-hydraulic brake can independently regulate each wheel's braking force and control the vehicle braking process in real-time, which is easy to realize the demand of vehicle intelligent driving. However, it should be pointed out that as a transitional product from traditional hydraulic brake to line-controlled brake, electro-hydraulic brake still needs complex braking pipelines to connect the pressure regulating unit and friction brake, there is still a risk of brake fluid leakage in electro-hydraulic brake; At the same time, the precise control of the braking force of a single wheel by electro-hydraulic brake depends on high-precision hydraulic components and high system working pressure, which increases the manufacturing cost and safety risk of electro-hydraulic brake.

## 1.4 Electro-Pneumatic Brake

In order to adapt to the intelligent requirements of active safety, intelligent braking and automatic driving, electro-pneumatic brake has become the inevitable development of pneumatic brake system. The electro-pneumatic brake is a new brake system, which can coordinate with the motor regenerative braking force and independently control the front and rear axle air suppression force. Compared with the parallel braking force distribution regenerative braking system has obvious advantages, not only improves the braking stability of the vehicle, but also can prolong the driving range.

As the core part of the electro-pneumatic brake, the response characteristics of the electronic pneumatic brake valve seriously affect the performance of the whole braking system and the safety of the vehicle. In the electro-pneumatic brake system, the electro-pneumatic brake valve is responsible for regulating the brake chamber pressure on the left and right sides of a single axle, and its independent electronic control unit is built-in. The working principle of the electro-pneumatic brake valve is as follows: when the ECU receives the brake signal, the compressed air provided by the air source enters the brake valve along the air pipe, the brake valve outputs the adjusted pressure to the brake air chamber, drives the piston of the brake air chamber to move, and the piston rod presses the brake disc to brake the wheel. The electro-pneumatic brake valve can realize the functions of pressurizing, reducing and

**Fig. 1.4** Appearance of electro-pneumatic brake valve

maintaining the pressure of the brake chamber, corresponding to the implementation, maintenance and release of braking of the vehicle. The appearance of the new generation front axle electro-pneumatic brake valve is shown in Fig. 1.4a [7].

As a new type of brake system, electro-pneumatic brake has incomparable advantages in braking force distribution and coordination with regenerative brake system. The electro-pneumatic brake solves the response delay caused by the long air passage of the traditional pneumatic brake system, and it is able to improve the performance and safety of pneumatic brake by improving the brake response speed.

## 1.5 Composite Brake System for Motor and Friction

Since the first generation, Toyota's hybrid vehicle Prius has used motor regenerative brake to recover braking energy. However, motor regenerative brake only acts on the driving wheel, and the output braking torque is small, which can not fully meet the braking needs of the driver. Therefore, motor regenerative and hydraulic brake must be used simultaneously to form a composite brake system for motor and friction.

As shown in Fig. 1.5 [8], composite brake system for motor and friction is mainly constituted by a driving motor, motor controller, power battery, battery management system, brake pedal simulator, pressure control unit, electro-hydraulic brake controller, and the whole vehicle controller.

As Fig. 1.6 [9] shows, the brake treadle simulator can collect the information of the brake treadle displacement sensor and the brake master cylinder pressure sensor in real-time and transmit it to the electro-hydraulic brake controller. The electro-hydraulic brake controller will identify the driver's braking intention according to the transmitted sensor information. If in the normal braking state, the electro-hydraulic brake controller sends the braking torque required by the driver to the

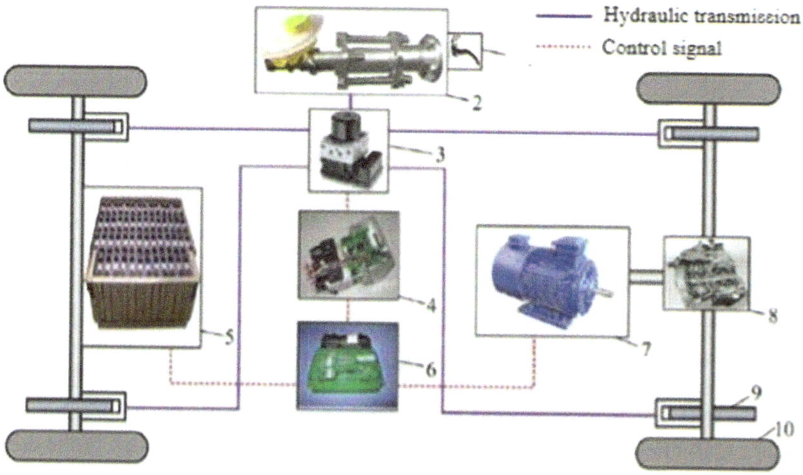


Where: 1. Brake pedal 2. Brake pedal Simulator 3. The pressure control unit 4. Electro-hydraulic brake controller 5. Power battery and battery management system 6. The whole vehicle controller 7. Drive motor and motor controller 8. Rear-axle 9. Friction brake 10. Wheel

**Fig. 1.5** Schematic diagram of motor and hydraulic hybrid brake system

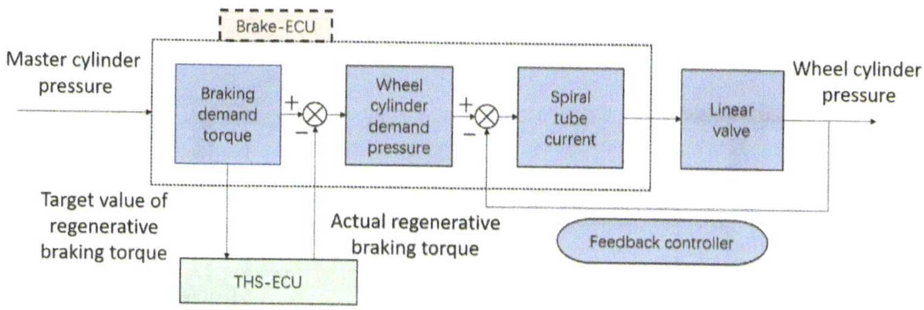


**Fig. 1.6** Control block diagram of motor and hydraulic hybrid brake system

whole vehicle controller. The whole vehicle controller computes the target braking torque of motor regenerative braking according to the driver's required braking torque, and vehicle status information and respectively sends it to the motor controller and electro-hydraulic brake controller; The difference between the driver's required braking torque and the motor regenerative brake target braking torque is the electro-hydraulic brake target braking torque. The electro-hydraulic brake controller controls the electro-hydraulic braking torque of the front and rear wheels through the high-speed solenoid valve on the basis of the braking force distribution method of the front and rear axles. In emergency braking, the whole vehicle controller will prohibit the motor regenerative braking and fully use the electro-hydraulic brake for anti-lock braking control.

## 1.6 Integrated Brake System for Electromagnetic and Friction

As a friction braking method, the braking process still relies on the friction between the brake lining and the brake disc to provide braking strength. When the car is driving on crowded city roads, the use of electromagnetic brake instead of electronic, hydraulic brake to achieve the driver's low-strength braking needs can effectively reduce the frequency of friction braking, significantly extend the service life of friction linings and brake discs, at the same time reduce the friction brake caused by brake dust and brake noise pollution. When the car is in the emergency braking state, because the electro-hydraulic brake precise control of a single wheel braking cost is high and the risk is greater, the use of electronic, hydraulic brake, and electromagnetic brake coordination work can significantly improve the response speed of the car emergency braking, enhance the accuracy of the car anti-lock brake control, effectively guarantee the safety of the car brake.

As shown in Fig. 1.7 [10], in Professor Anwar S's proposal for integrated brake system for electromagnetic and friction, electronic, hydraulic brake, and electromagnetic brake are only functional integrations; Integrated brake system for electromagnetic and friction includes integrated brake units (friction brakes, electromagnetic brakes), electric drive units (battery, rectifiers), hydraulic units (electronic brake

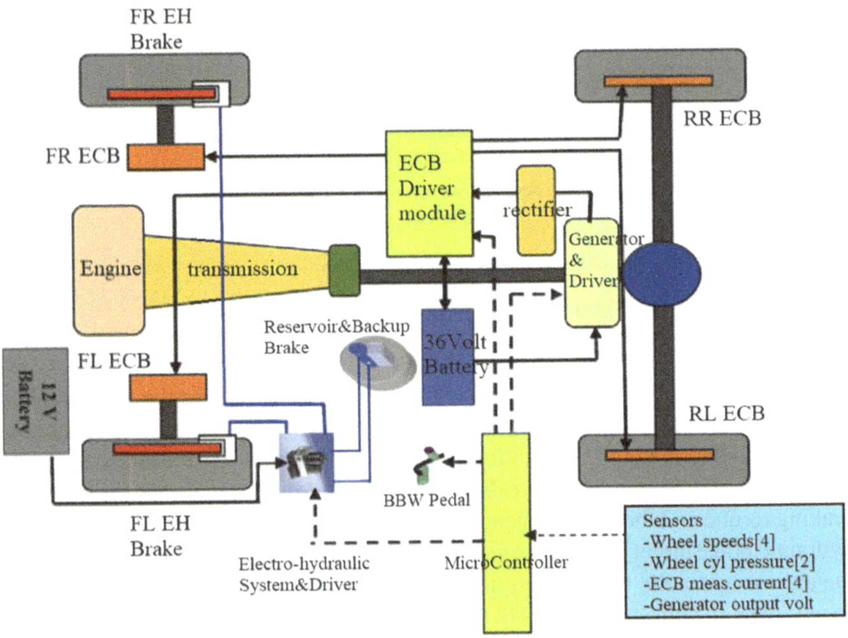


**Fig. 1.7** Structure diagram of integrated brake system for electromagnetic and friction

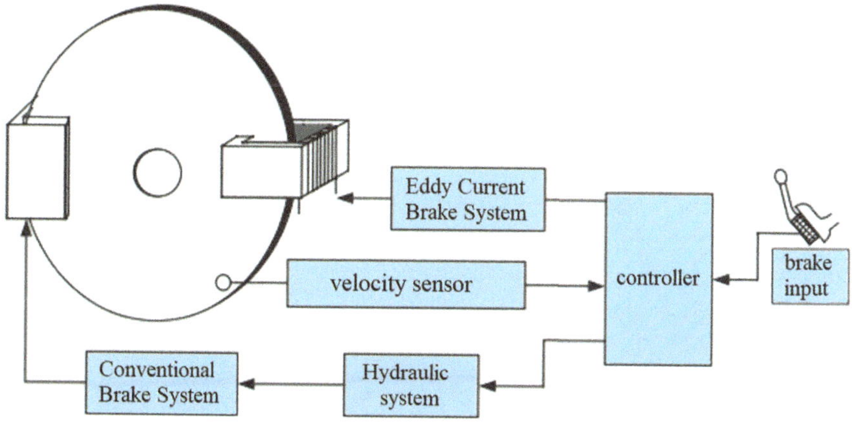


**Fig. 1.8** System integration program for electromagnetic brake and friction brake proposed by professor song

pedals, brake master cylinders, brake lines), control units (electro-hydraulic brake control units, electromagnetic brake control units, microprocessors, etc.).

As Fig. 1.8 [11] shows, in Professor Song's proposal for integrated brake system for electromagnetic and friction, the electromagnetic brake discs are combined with the friction brake discs for functional and structural integration, which overcomes the vehicle wheel mounting space limitations.

## References

1. Jingming Z, Baoyu S, Xiaojing N (2008) Optimization of parallel regenerative braking control strategy [C]. IEEE
2. Lee Jr K, Park K (2001) Modeling of the eddy currents with the consideration of the induced magnetic flux [C]. IEEE
3. Reuter DF, Lloyd EW, Zehnder JW, et al (2003) Hydraulic design considerations for EHB systems [R]. SAE Tech Paper
4. Morey B (2012) Continental focuses on faster braking with electrohydraulic system [J]. Automot Eng Int 120(3)
5. Wire control electronic hydraulic brake system [EB/OL]. http://www.tongyuauto.com/info/70130.html
6. Bosch Brake-by-Wire System iBooster [EB/OL]. http://mp.ofweek.com/nev/a545673526246
7. Quan'an H, Qian P (2020) Modeling and experimental verification of electronically controlled pneumatic brake valve for commercial vehicles [J]. Times Automobile 05:100–104
8. Kuiyang W, He R (2014) Analysis on the current situation of coordinated control technology of automobile electromechanical composite braking system [J]. J Chongqing Univ Technol (Nat Sci) 12:10–17
9. Sakai A, Sasaki Y, Otomo A, et al (1997) Toyota braking system for hybrid vehicle with regenerative system [J]. EVS-14, Orlando, Florida, USA

10. Anwar S (2005) Generalized predictive control of yaw dynamics of a hybrid brake-by-wire equipped vehicle [J]. Mechatronics 15(9):1089–1108
11. Song J (2005) Performance evaluation of a hybrid electric brake system with a sliding mode controller [J]. Mechatronics 15(3):339–358

# Chapter 2
# Research Status of Hybrid Brake-By-Wire System

In order to guarantee the stability of vehicle braking, in the hybrid brake-by-wire system, a reasonable braking force distribution method of vehicle front and rear axles and mode switching strategy under coordinated braking must be adopted. This chapter introduces the research status of braking force distribution method and mode switching of intelligent vehicle hybrid line control system.

## 2.1 Braking Force Distribution Control of Hybrid Brake-By-Wire System

The factors that need to be considered in the braking force distribution of the front and rear axles of the automobile integrated brake system are complex, including the national braking regulations, the performance of the integrated brake system and the restrictions of the vehicle running state. The braking force distribution methods of the front and rear axles of the automobile integrated brake system are summarized below.

The braking force distribution methods of automobile front and rear axles based on composite brake system for motor and friction can become roughly divided into four types: simple superposition type, parallel distribution type, ideal braking force distribution type and energy recovery maximization type. Except for the simple superposition braking force distribution method (Fig. 2.1a [1]), the other three distribution methods need to redistribute the braking force for motor regenerative and hydraulic brake (Fig. 2.1b [1]).

The simple superposition braking force distribution method doesn't change the braking force distribution of the front and rear axles of the traditional hydraulic brake, but merely the simple superposition of the braking torque of the motor regenerative brake and the traditional hydraulic brake. This braking force distribution method has strong applicability, but its braking energy recovery efficiency is low. Due to the



D. Hu, *Design and Control of Hybrid Brake-by-Wire System for Autonomous Vehicle*, Key Technologies on New Energy Vehicles,
https://doi.org/10.1007/978-981-16-8946-8_2

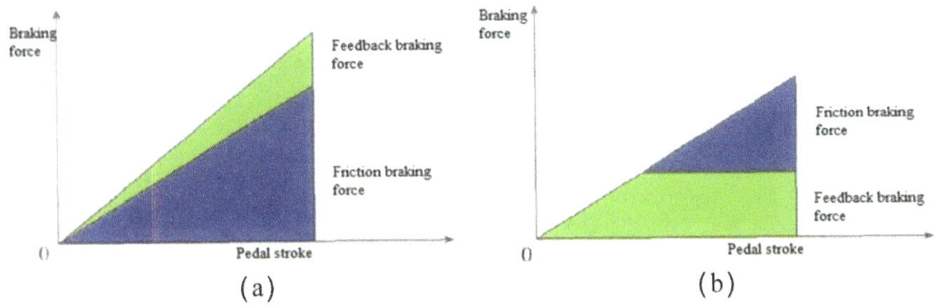


**Fig. 2.1** Schematic diagram for the braking force distribution of motor and hydraulic hybrid brake system

high vehicle speed, to guarantee the safety of braking, changing the braking force distribution of front and rear axles of traditional hydraulic brake will reduce the safety redundancy. Therefore, new energy vehicle enterprises represented by Tesla, BMW, BYD and SAIC generally adopt the simple superposition braking force distribution method. On this basis, some European and American vehicle enterprises put forward the operation mode of "single-pedal driving". The vehicle enters the sliding mode when the driver releases the accelerator pedal. That is, the motor regenerative brake begins to be implemented. The operation mode of "single-pedal driving" can reduce the frequency of the driver stepping on the brake pedal and use motor regenerative brake to recycle vehicle braking energy as much as possible before hydraulic brake.

The parallel distributed braking force distribution method requires that when the required braking strength of the driver is less than the set lower limit, only the motor regenerative brake is used, and the hydraulic brake does not work; When the driver's required braking strength is more than the set upper limit, the vehicle enters the emergency braking state, the motor regenerative brake stops working, and the hydraulic brake implements anti-lock braking control; When the driver required braking strength is between the lower limit and the upper limit, the braking force on the driving wheel is jointly provided by the motor brake force and hydraulic brake. The braking force on the non-driving wheel is separately provided by hydraulic brake, as Fig. 2.2b [2] shows. Similar to the simple superposition braking force distribution method, the parallel distribution braking force distribution method only needs to control the motor regenerative brake without active adjustment of hydraulic brake. Due to frequent braking and low braking intensity under urban bus working conditions, new energy vehicle enterprises represented by Zhengzhou Yutong and Suzhou Jinlong adopt parallel distribution braking force distribution method. The parallel distributed braking force distribution method proposed by Guozhu [3] takes the braking force applied by the motor regenerative brake on the driveshaft as the main control means. It doesn't change the braking force distribution of the front and rear axles of the traditional hydraulic brake. The research results show that the active control of the motor regenerative brake can make sure the vehicle's braking stability when it is on the high adhesion road, but at this time, the recovery effect of vehicle braking energy is poor.

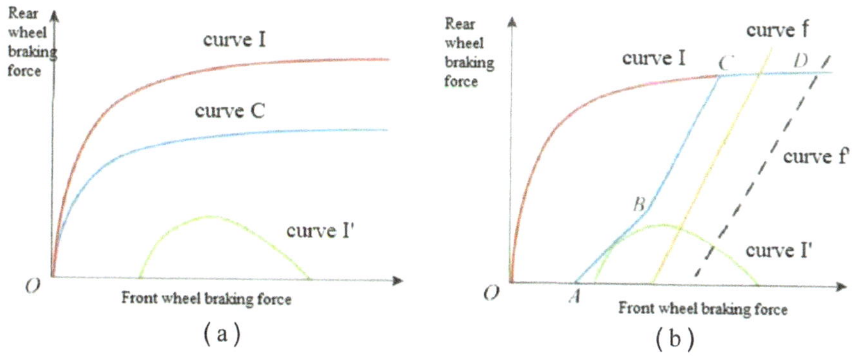


**Fig. 2.2** Schematic diagram for the distribution of vehicles' front and rear braking force

The ideal braking force distribution method requires that the braking force of the front and rear axles of the vehicle shall always be distributed in line with the ideal braking force distribution curve (i.e., "curve") so that the vehicle can make better use of the ground adhesion coefficient and prevent the wheels of the front and rear axles from locking in advance, as Fig. 2.2a [2] shows. Simultaneously, the vehicle controller also needs to consider the external characteristics of the motor and the state of the power battery to determine the braking force distribution of hydraulic brake and motor brake on the driveshaft. Kim et al. [2] used the same wheel slip ratio of the rear axle as the control target and used the fuzzy control method to control the braking force applied by the composite brake system for motor and friction on the front and rear axles. Meanwhile, the braking force of the front and rear axles is distributed according to the ideal braking force distribution curve. The research results show that the ideal braking force distribution method can significantly improve the braking stability of vehicles.

The energy recovery maximization braking force distribution method requires that only the motor regenerative brake is used when the driver required braking strength is less than the set value. When the driver required braking strength is greater than the set value, under the condition that the front and rear axles of the vehicle are not locked, the braking force distribution of the front and rear axles is optimized to maximize the motor regenerative braking energy. For the 4WD electric vehicle equipped with a composite brake system for motor and friction, Chaohe [4] puts forward the energy recovery maximization braking force distribution method, which prioritizes the use of motor regenerative brake and uses hydraulic braking force to meet the residual braking demand of the vehicle. The research results show that the energy recovery maximization braking force distribution method can ensure the braking stability on the premise of ensuring the braking stability. The energy recovery efficiency of the automobile is greatly improved.

## 2.2 Mode Switching Control of Hybrid Brake-By-Wire System

According to the classification of braking conditions, the coordinated control of automobile integrated braking system can be divided into two classifications, and the first classification is the coordinated control for ordinary brake, which is mainly through the coordinated control of automobile integrated brake system to make sure the stability of braking mode switching and the comfort of the braking process. The second type is the coordinated control for emergency braking conditions, which is mainly through the braking force applied on the wheels by the vehicle integrated brake system to prevent wheel locking and make sure the safety of vehicle emergency braking. Next, the coordinated control methods of automobile integrated brake system under two braking conditions are summarized.

### *2.2.1 Mode Switching Control Under Non-emergency Braking*

(1) Braking mode switching stability

A kind of complex system including discrete event dynamic system and continuous variable dynamic system is called hybrid dynamic system (HDS). Its state evolution process is a hybrid motion process. At present, the hybrid dynamic system theory has been applied and popularized in automobiles. Borrelli et al. [5] applied the hybrid logic dynamic model to study the problem of vehicle traction control, so as to achieve the purpose of maximum traction on wet and slippery roads and improve the driver handling ability of the vehicle. Pettersson et al. [6] established the hybrid characteristic modeling of the working process of the automatic transmission, solved the piecewise quadratic Lyapunov function that makes the system stable by using the method of linear matrix inequality and achieved good application results. Aiming at the nonlinear problem of automobile electronic throttle, Baotic et al. [7] used a piecewise affine model to model it and carried out nonlinear control on this basis. Chunhua [8] and others use the switching system model to model the energy management system of the hybrid electric vehicle and use the logical judgment method to establish the mapping set. The analysis results show that this method is feasible.

In practical application, the working process of the vehicle integrated brake system includes the process of discrete state variables and continuous state variables, and the discrete variables and continuous variables interact and restrict each other. The discrete state variables include the braking mode switching control command, etc., while the continuous state variables include the vehicle speed, acceleration, brake wheel cylinder pressure, etc. Therefore, the vehicle integrated brake system can be regarded as a hybrid dynamic system. It should be noted that stability is the most basic performance of the system. Due to the introduction of "switching", the stability of the vehicle integrated brake system has changed greatly. For example, although each braking mode can work stably, the improper switching control law will lead to the instability of the whole integrated brake system. For another example, although the integrated brake system is unstable in each braking mode, the entire integrated brake system may tend to work stably through appropriate switching control law.

Wenyan [9] applies the hybrid dynamic theory to define the identification of discrete events and continuous variables in the vehicle braking process, then describes and analyzes the hybrid dynamic behavior of composite brake system for motor and friction, establishes the differential Petri model of braking mode switching, and puts forward the stability control method of composite brake system for motor and friction in the braking process. At present, when establishing the control model of integrated brake system for electromagnetic and friction, researchers at home and abroad generally only consider the influence of discrete variables or continuous variables, ignoring the interaction and influence of the two variables. At the same time, they do not involve the control strategy and stability analysis of integrated brake system for electromagnetic and friction in the braking mode switching process.

(2) Comfort during braking

On the basis of the response characteristics of electro-hydraulic brake and electromagnetic brake, there is no relevant research work at home and abroad to coordinate and control electro-hydraulic brake and electromagnetic brake to improve vehicle braking comfort and braking feeling consistency. However, in the research field of composite brake system for motor and friction, the research results in this direction have been very fruitful. The coordinated control methods proposed by domestic and foreign researchers mainly include the response characteristic correction method, filter balance method and double closed-loop control method.

In order to solve the difference in dynamic response characteristics between motor regenerative brake and hydraulic brake, the response characteristic correction method first needs to analyze the dynamic characteristics of hydraulic brake and motor regenerative brake, then simplify the dynamic characteristic model of hydraulic brake and motor regenerative brake, and finally correct the dynamic response characteristics of motor regenerative brake and hydraulic brake. Numasato et al. [10] found that the response characteristic correction method can make the hydraulic brake and motor regenerative brake have the same response speed and coordinate and control the hydraulic brake and motor regenerative brake accordingly.

In the process of coordinated control using the filter balance method, the motor regenerative brake is required to meet the high-frequency part of the driver braking demand, while the electro-hydraulic brake is required to complete the low-frequency part of the driver braking demand. The filtering algorithm is used to filter the respective target braking torque to balance the dynamic response characteristics of the two. However, Okano et al. [11] found that when the driver demand for braking intensity is low, using hydraulic brake to meet the demand of low-frequency braking torque will lead to low efficiency of vehicle braking energy recovery.

When the feed-forward control method is used for the coordinated control of hydraulic brake and motor regenerative brake, the target braking torque of hydraulic brake is the difference between the driver-required braking torque and the motor regenerative brake torque. However, due to the difficulty of accurate control of hydraulic brake, the driver-required braking torque fluctuates, and the braking comfort becomes worse in the braking process. The feedback control method takes the fluctuation of the driver-required braking torque as the control target and implements the compensation control through the motor regenerative brake to improve the comfort of the braking process. However, compared with the feed-forward control method, the motor regenerative braking intensity in the feedback control method is lower, so the braking energy recovery efficiency is also lower. The double closed-loop control method is developed based on the feed-forward control method and feedback control method, taking into account the advantages of the feed-forward control method and feedback control method. The double closed-loop control method proposed by Sakamoto et al. [12] can fully implement motor brake to improve energy recovery efficiency on the premise of maintaining the fluctuation stability of driver required braking torque.

## 2.2.2 Mode Switching Control Under Emergency Braking

Because the response characteristics and control performance of motor regenerative brake and electromagnetic brake are better than hydraulic brake, in case of emergency, composite brake system for motor and friction needs to implement the coordinated control of motor regenerative brake and hydraulic brake to ensure the braking safety of vehicles. At present, there are three coordination methods of composite brake system for motor and friction under emergency braking: hydraulic active control method, motor active control method and motor hydraulic coordinated control method.

The hydraulic active control method requires that when the wheel tends to lock, the motor regenerative brake stops working, and the hydraulic brake implements anti-lock braking control. However, when the coordinated control is carried out according to the hydraulic active control type, because the response time of hydraulic brake is slower than that of motor regenerative brake, if the motor regenerative brake exits quickly and the hydraulic pressure rise is slow, it will lead to the problem of exiting the hydraulic anti-lock braking control because the intervention conditions of hydraulic anti-lock braking are not met at this time. For the commercialized new energy vehicles, it is required to take the control mode of motor regenerative brake force withdrawal under emergency braking conditions to avoid the impact of motor regenerative brake on the normal operation of the original mature hydraulic anti-lock braking control system.

When it is detected that the front and rear axle wheels are locked, the motor active control method can effectively improve the ride comfort of anti-lock braking by controlling the regenerative brake force of the motor to realize the anti-lock function of the vehicle; However, the regenerative braking force provided by the motor is limited, and the braking force demand is great during emergency braking. It is difficult to meet the braking demand of the driver simply by relying on the

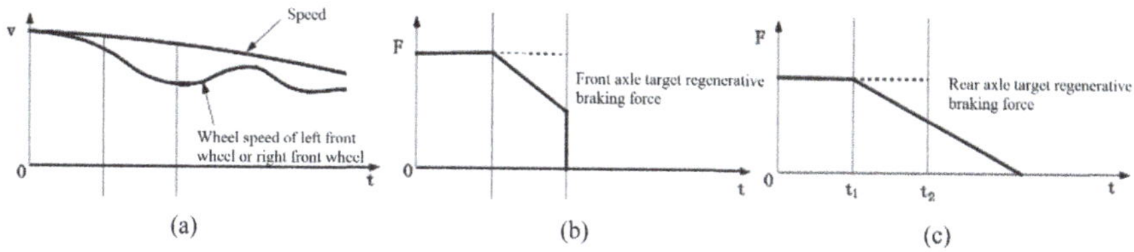


**Fig. 2.3** Schematic diagram for the control of motor brake force in front and rear axis under emergency braking

regenerative braking output of the motor. Therefore, the active control method of the motor still needs hydraulic brake to provide part of the braking force. As Fig. 2.3 [13] shows, the anti-lock braking control state starts at this moment, and the regenerative braking torque of front and rear axle motors gradually decreases according to a certain slope; When the time is reached because the locking trend of the front axle is not relieved, the regenerative braking force of the front axle motor is directly reduced to zero, while the regenerative braking force of the rear axle motor continues to decrease according to a constant slope.

The motor and hydraulic coordinated control method requires to give full play to the respective superiorities of hydraulic brake and motor regenerative brake and improve the control performance of automobile anti-lock braking by coordinating the control of the hydraulic braking force and motor braking force. As shown in Fig. 2.4 Lei et al. [14, 15], proposed that the motor and hydraulic coordinated anti-lock braking control is composed of the upper control module, middle control module and lower control module. The upper control module includes driver braking intention recognition function, road state recognition function and slip rate distribution function of the front and rear wheel; The middle control module includes the front and rear wheel slip rate control function; The bottom control module includes motor control function and front and rear wheel hydraulic braking control function.

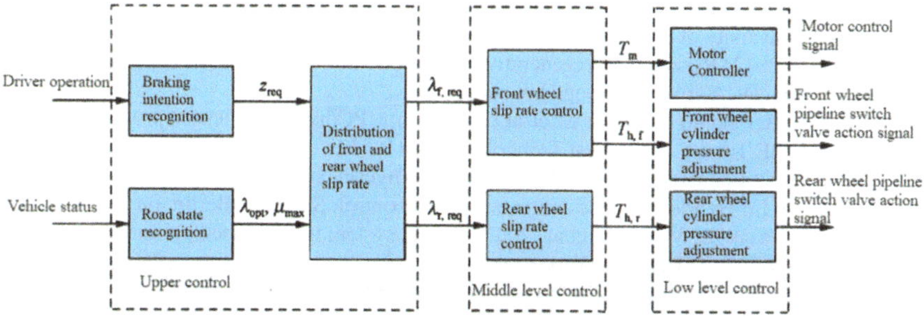


**Fig. 2.4** Block diagram of the coordinated control based on slip rate assignment for the anti-lock braking

Taking advantage of the advantages of convenient and reliable electromagnetic braking, fast response speed, and high control accuracy, Xuejun [16] studied the real-time observation and estimation algorithm of pavement state to realize the accurate identification of pavement condition by the controller of integrated brake system for electromagnetic and friction. Meanwhile, to consider the nonlinear characteristics of integrated brake system for electromagnetic and friction, and take the optimal slip rate as the Control objectives, the anti-lock braking controller of integrated brake system for electromagnetic and friction is designed. The anti-lock braking performance of single electro-hydraulic brake and integrated brake system is studied under different road conditions. Verify the advantages of integrated brake system for electromagnetic and friction in reducing emergency braking distance. However, the anti-lock braking control method of integrated brake system for electromagnetic and friction proposed by Xuejun [17] does not coordinate the control of electro-hydraulic brake and electromagnetic brake, nor does it quantify the coordinated control process of the two braking modes. Because the dynamic response characteristics of electro-hydraulic brake and electromagnetic brake are different, and the output characteristics of electromagnetic braking torque are changing with the vehicle driving conditions and electro-hydraulic braking force, how to give full play to the superiorities of electro-hydraulic brake and electromagnetic brake, and how to coordinate and control the two methods to improve the anti-lock braking performance of the vehicle is worthy of in-depth analysis and research.

## References


1. Junzhi Z, Chen LV, Yutong L (2014) Hybrid drive and hybrid braking system of electric vehicles—current situation and prospect. J Automot Safety Energy Saving 5(03):209–223
2. Kim D, Kim H (2006) Vehicle stability control with regenerative braking and electronic brake force distribution for a four-wheel drive hybrid electric vehicle[J]. Proceedings of the Institution of Mechanical Engineers, Part D: Journal of Automobile Engineering 220(6):683–693
3. Guozhu Z (2012) Research on several key issues of regenerative braking for electric vehicles. Nanjing University of Aeronautics and Astronautics
4. Chaohe T (2012) Research on regenerative braking control strategy of in-wheel motor driven electric vehicles. Beijing Jiaotong University
5. Borrelli F, Bemporad A, Fodor M et al (2006) An MPC/hybrid system approach to traction control. IEEE Trans Control Syst Technol 14(3):541–552
6. Pettersson S, Lennartson B (2000) Stability of hybrid systems using LMIs—a gear-box application. In: Hybrid systems: computation and control. Springer, Berlin, pp 381–395
7. Baotic M, Vasak M, Morari M et al (2003) Hybrid system theory based optimal control of an electronic throttle. IEEE, New York
8. Chunhua H, Ren H, Nan L (2011) Optimal design of HEV energy control system based on switching system. J Jiangsu Univ (Nat Sci Edition) (05):497–501
9. Wenyan C (2010) Stability analysis of automobile regenerative braking based on hybrid system theory. Jiangsu University
10. Numasato H, Tomizuka M (2003) Settling control and performance of a dual-actuator system for hard disk drives. IEEE/ASME Trans Mechatron 8(4):431–438
11. Okano T, Sakai S, Uchida T et al (2002) Braking performance improvement for hybrid electric vehicle based on electric motor's quick torque response. EVS19

12. Sakamoto T, Hirukawa K, Ohmae T (2006) Cooperative control of full electric braking system with independently driven four wheels. IEEE, New York
13. Hara M, Nihei T, Matsui N et al (2005) Braking force control apparatus for a vehicle. Google Patents
14. Lei Z, Yugong L, Keqiang L et al (2009) Integrated control of feedback braking and anti lock braking of electric vehicles. J Tsinghua Univ (Nat Sci Edition) (05):728–732
15. Lei Z, Yugong L, Diange Y et al (2008) Electric vehicle braking control strategy based on slip ratio test. J Tsinghua Univ (Nat Sci Edition) 48(5):883–887
16. Xuejun L, Ren H (2014) Electromagnetic-hydraulic compound anti-lock braking system sliding mode control. Trans Chin Soc Agric Mach 45(5):1–7
17. Jun S, Ren H, Xuejun L (2013) The electromagnetic retarder braking torque calculation based on the force model in the magnetic field of a moving conductor. Automobile Technol (03):29–32

# Chapter 3
# Modeling of Hybrid Brake-By-Wire System

Both electro-hydraulic brake and electromagnetic brake have nonlinear characteristics. This chapter establishes the dynamic mathematical models of electro-hydraulic brake, electromagnetic brake respectively, and deduces the coupling nonlinear model of integrated brake system for electromagnetic and friction. This chapter also establishes the electromagnetic brake model, electro-pneumatic brake model and model of composite brake system for motor and friction.

## 3.1 Motor Brake Model

### *3.1.1 Driving Motor Model*

Braking energy recovery is closely related to the performance of driving motor. On the one hand, the more power a driving motor can generate, the more energy it can theoretically recover. On the other hand, the more braking force a drive motor can provide, the more energy it can provide for recovery. The modeling in this section refers to the permanent magnet synchronous motor model. Generally, researchers use the *d*-*q* equation to study motor characteristics. Its principle is to establish a synchronous rotation coordinate system on the rotor of the motor, set axis $d$ and axis $q$ as the magnetic field direction of the rotor and its vertical direction respectively. In order to obtain good control characteristics, the mathematical model of driving motor can be studied in this coordinate system.

The state equation of the *d*-*q* axis driving motor model can be expressed in Formula (3.1):



D. Hu, *Design and Control of Hybrid Brake-by-Wire System for Autonomous Vehicle*, Key Technologies on New Energy Vehicles,
https://doi.org/10.1007/978-981-16-8946-8_3

$$\begin{cases} \dot{i}_d = -\frac{R_S}{L} i_d + P\omega_m i_q + \frac{1}{L} u_d \\ \dot{i}_q = -\frac{R_S}{L} i_q - P\omega_m i_d + \frac{1}{L} u_d - P\frac{\psi}{L}\omega_m \\ J_m \dot{\omega}_m = T_b^m - T_m - B_m \omega_m \end{cases} \tag{3.1}$$

where: $i_d$ and $i_q$—the current of $d$-$q$ axis, A; $u_d$ and $u_d$—the voltage of the $d$-$q$ axis, V; $R_s$—resistance, Ω; $L$—inductance, H; $P$—the number of magnetic poles; $\psi$—magnetic flux, Wb; $\omega_m$—the angular velocity of the motor, rad/min; $J_m$—the moment of inertia of the motor; $B_m$—viscous damping coefficient; $T_m$—motor torque, N m.

The driving motor torque can be expressed in Formula (3.2):

$$T_m = \frac{3}{2} P \psi i_q \tag{3.2}$$

The power of the driving motor can be expressed in Formula (3.3):

$$\begin{cases} \frac{T_m \omega_m}{\eta_m}, motor \\ T_m \omega_m \eta_m, generator \end{cases} \tag{3.3}$$

where: $\eta_m$—the efficiency corresponding to the angular velocity $\omega_m$ when the output torque of the driving motor is $T_m$.

The efficiency of the driving motor is related to the speed and torque. The efficiency of the driving motor is obtained through experiments, as shown in Fig. 3.1. In this section, eddy current and hysteresis loss in the motor are ignored when modeling. Eddy current will increase the temperature of the iron core and cause power loss. Hysteresis loss will increase the temperature of the motor, resulting in a decrease in efficiency.

### 3.1.2 Energy Storage Model

#### 3.1.2.1 Power Battery

PEMFC has the advantage of high energy conversion efficiency, and is the main power source for converting chemical energy from hydrogen to electrical energy for vehicle operation. Analyzing the dynamic characteristics of PEMFC, the efficiency of PEMFC is defined as:

$$\eta_{fc} = \frac{P_{fc\_act}}{Q_{LHV}^{H_2} W_{H_2}}. \tag{3.4}$$

where: $P_{fc\_act}$ is the actual output power of PEMFC, $Q^{H_2}{}_{LHV}$ is the low heating value of hydrogen, and $W_{H_2}$ is the hydrogen consumption, which can be expressed as:

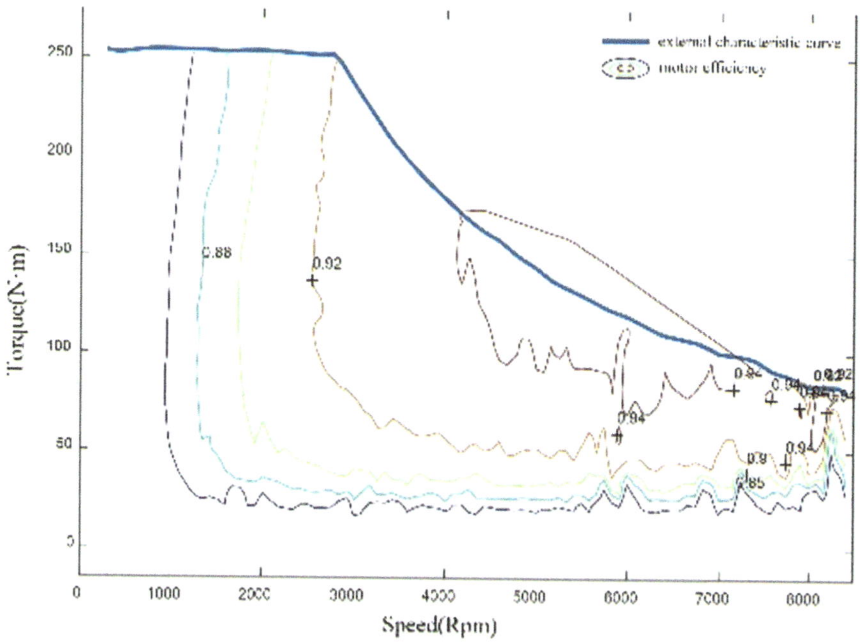


**Fig. 3.1** Drive motor efficiency characteristics

$$W_{H_2} = \frac{I_{st} n_{cell} M_{H_2}}{2F}. \tag{3.5}$$

where: $I_{st}$ is the current of the PEMFC stack, which can be calculated by $P_{fc\_act}$, $n_{cell}$ is the number of individual cells in the stack, $M_{H_2}$ is the molar mass of hydrogen, and $F$ is the Faraday constant.

This paper establishes a fuel cell vehicle dynamic characteristics model considering the dynamic characteristics of PEMFC. For the fuel cell vehicle model, PEMFC will be expressed as a non-minimum phase first-order linear time-invariant (LTI) system, and its state space is expressed as:

$$\begin{cases} \dot{P}_{fc\_n}(t) = AP_{fc\_n}(t) + BP_{fc\_req}(t) \\ P_{fc\_act}(t) = CP_{fc\_n}(t) + DP_{fc\_req}(t) \end{cases}. \tag{3.6}$$

where: $P_{fc\_req}(t)$ is the required power of PEMFC at the time as an input, $P_{fc\_act}(t)$ is the power delivered by FCS to the power system at time t as output, $P_{fc\_n}(t)$ is the dynamic state of PEMFC and A, B, C and D are the states, input, output and feedthrough matrix respectively.

#### 3.1.2.2 Fuel Cell

This article chooses the widely used internal resistance model. By equating the power battery as a circuit in which a resistor is connected in series with an ideal voltage source, the model can be expressed as:

$$V_{bat} = V_{oc} - I_{bat} \cdot R \tag{3.7}$$

where: $V_{bat}$ is the battery terminal voltage, $V_{oc}$ is the open-circuit voltage of the battery, $R$ is the internal resistance of the battery, and $I_{bat}$ is the battery current, which can be expressed as:

$$I_{bat} = \frac{V_{oc} - \sqrt{V_{oc}^2 - 4R \cdot P_{bat}}}{2R}. \tag{3.8}$$

where: $P_{bat}$ is battery power, which can be expressed as:

$$P_{bat} = P_{mot,bat} + P_{acc}. \tag{3.9}$$

where: $Pmot$,bat is the power interaction between the power battery and the drive motor; $Pacc$ is the power of the electric accessory, which is set to a fixed average accessory power value during the running of the vehicle. The battery SOC is used to reflect the remaining capacity of the battery, which is defined as the ratio of the remaining capacity Q of the battery to the maximum capacity QC:

$$SOC = \frac{Q}{Q_C}. \tag{3.10}$$

SOC is calculated by the ampere-hour integral method:

$$SOC = \begin{cases} \frac{1}{Q_C}(Q_{\text{int}} - \frac{1}{3600}\int_0^t I_{bat}\text{d}t);\ I \geq 0 \\ \frac{1}{Q_C}(Q_{\text{int}} - \frac{1}{3600}\int_0^t \eta_{bat} \cdot I_{bat}\text{d}t);\ I < 0 \end{cases}. \tag{3.11}$$

where: $Qint$ is the initial capacity of the battery, and $\eta bat$ is the coulombic efficiency of battery charging.

## 3.2 Electromagnetic Brake Model

### 3.2.1 *Nonlinear Analysis of Soft Magnetic Materials*

The main material of the brake disc is gray cast iron, while the material of the iron core is generally industrial pure iron. Both materials are soft magnetic materials,

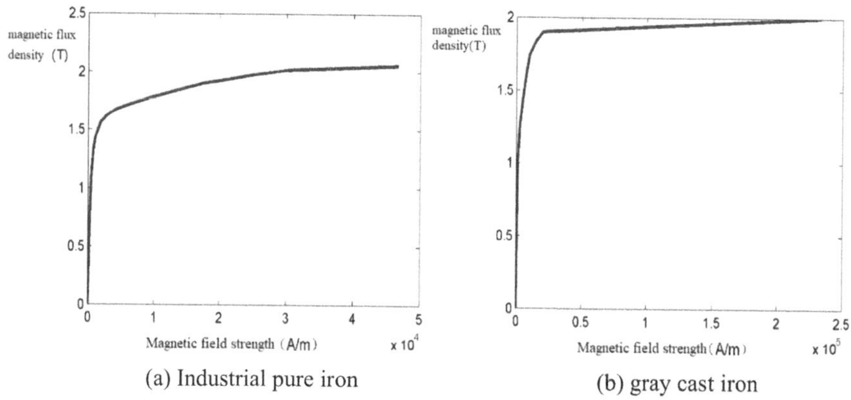


**Fig. 3.2** Magnetization curves of the soft magnetic materials

and the variation curve of magnetic induction intensity with the applied magnetic field intensity is saturated and nonlinear. In order to meet the demands of accurately controlling electromagnetic braking torque, the nonlinear characteristics of soft magnetic materials need to be investigated. The magnetization curves of soft magnetic materials are mostly obtained by actual measurement, and the saturation nonlinear characteristics are usually expressed by piecewise linearization, as shown in Fig. 3.2 Using a power series expressed as a function, and the coefficient determined by the least square method, the functional relationship of the magnetization curve of soft magnetic material is obtained as follows:

$$\begin{cases} H_i = f_1(B_i) = c_1 B_i + c_2 B_i^3 \\ H_d = f_2(B_d) = c_3 B_d + c_4 B_d^3 \end{cases} \tag{3.12}$$

where: *Hi*, *Hd*—magnetic field strength of the iron core and brake disc, A/m; *Bi*, *Bd*—Magnetic induction strength of the iron core and brake disc, T; *c1*, *c2*, *c3*, *c4*—Fitting coefficient.

During the braking process of the electromagnetic brake, the brake disc absorbs the vehicle kinetic energy and converts it into heat energy, which will be dissipated, and the excitation coil will also generate some heat when it is energized for a long time. Under the demand of accurately controlling the electromagnetic braking torque, the nonlinear characteristics of the resistivity and relative permeability of soft magnetic materials and the resistivity of copper conductor with temperature need to be investigated. As shown in Fig. 3.3a, the variation curve of relative permeability with temperature is processed by piecewise linearization:

$$\mu_r = \begin{cases} \mu_1 + a_1 t_1 \ 273 < t_1 < 473 \\ \mu_2 + a_2 t_1 \ t_1 \geq 473 \end{cases} \tag{3.13}$$

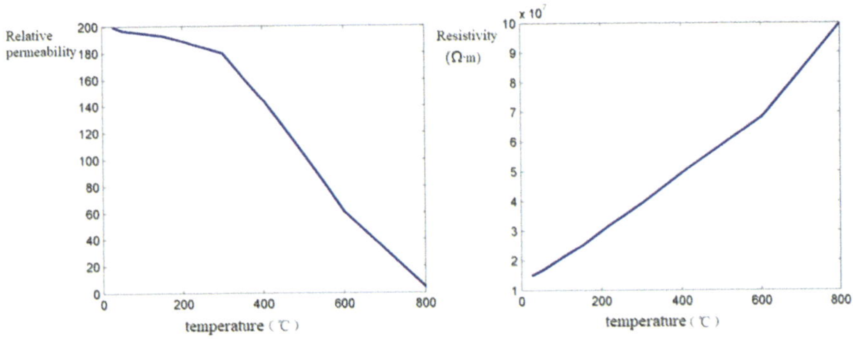


(a) Relative permeability (b) Resistivity

**Fig. 3.3** Curves of the brake disc's electromagnetic parameters changing with temperature

where: $\mu_1$, $\mu_2$—relative permeability constant; $a_1$, $a_2$—Temperature coefficient of brake disc permeability, 1/K; $t_1$—Brake disc temperature, K.

As shown in Fig. 3.3b, considering the strong heat dissipation capability of the disc brake and the narrow temperature variation extent of copper conductor, it is proposed to directly linearize the variation curve of resistivity with temperature:

$$\begin{cases} \rho_d = \rho_1[1 + b_1(t_1 - t_0)] \\ \rho_w = \rho_2[1 + b_2(t_2 - t_0)] \end{cases} \quad (3.14)$$

where: $\rho_w$—resistivity of copper conductor, Ω.m; $\rho_1$, $\rho_2$—Resistivity of brake disc and copper conductor at room temperature; $b_1$, $b_2$—Resistivity temperature coefficient of brake disc and copper conductor, 1/K; $t_0$—Ambient temperature, K; $t_2$—Copper conductor temperature, K.

### 3.2.2 *Calculation of Electromagnetic Braking Torque*

According to Faraday's law of electromagnetic induction, when the electromagnetic brake works, the eddy current magnetomotive force and excitation magnetomotive force are AC excitation, which makes the magnetic circuit analysis of the electromagnetic brake complex. Thus, assuming that the nonlinear mathematical model of electromagnetic brake is a steady-state model, the magnetic path of electromagnetic brake is equivalent to DC magnetic circuit. According to Kirchhoff's law of magnetic circuit and the constraints of the relationship between magnetic flux and magnetic pressure of each magnetic circuit: assuming that the excitation magnetic field, the cross-sectional area of air gap magnetic field and brake disc magnetic field is identical, the nonlinear equation for calculating magnetic flux is obtained:

$$NI - \frac{\sqrt{2}\phi\omega_d\Delta_h}{8\pi\rho_d} = \frac{\phi}{S_p\mu_0}l_g + f_1(\phi)\Delta_h + f_2(\phi)h \tag{3.15}$$

where: $N$—turns of excitation coil; $I$—Input excitation coil current, A.

The calculation formula of electromagnetic braking torque is:

$$T_e = k_{eb}\omega_d\phi^2 \tag{3.16}$$

Of which: $k_{eb} = \frac{N_p k_a S_p \Delta_h}{8\pi\rho_d}$

$$\Delta_h = \sqrt{\frac{2\rho_d}{\omega_d\mu_0\mu_r}}$$

where: $k_{eb}$—electromagnetic braking gain coefficient; $N_p$—Pole pairs; $k_a$—Effective area gain coefficient.

## 3.3 Electro-Hydraulic Brake Model

### *3.3.1 Airbag Accumulator*

As shown in Fig. 3.4, in electro-hydraulic brake, the accumulator function stores high-pressure brake fluid from the motor pump and provides high-pressure brake fluid to the friction brake. When the pressure at the inlet of the accumulator is more than the gas pressure in the accumulator, hyperbaric brake fluid will enter the accumulator. At this time, the fluid pressure balance equation in the connecting pipeline is:

$$P_P - P_A = R_A Q_A + L_A\frac{dQ_A}{dt} \tag{3.17}$$

where: $P_P$—working pressure of the hydraulic pump, Pa; $P_A$—Accumulator gas pressure, Pa; $Q_A$—Actual flow of accumulator, $m^3/s$; $R_A$—Hydraulic resistance of accumulator connecting pipeline, $N{\cdot}s/m^5$; $L_A$—Hydraulic sensing of accumulator connecting pipeline, $N{\cdot}s^2/m^5$.

The flow continuity equation in the connecting pipeline is:

$$Q_A = C_A\frac{dP_A}{dt} \tag{3.18}$$

where: $C_A$—liquid capacity of the accumulator, $m^3/N$.

Assuming that the flow of the accumulator is positive when charging and negative when discharging, the liquid capacity of the accumulator is:

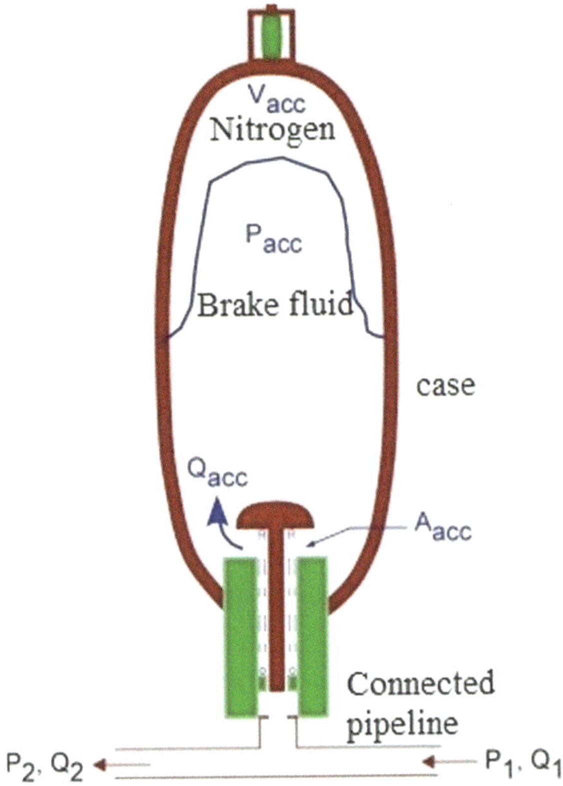


**Fig. 3.4** Schematic diagram of the accumulator's structure

$$C_A = \frac{V_{A0}}{nP_{A0}} \tag{3.19}$$

where: $V_{A0}$—volume of accumulator under initial state, m$^3$; $P_{A0}$—Pressure of accumulator under initial state, Pa; $n$—The polytropic process index of gas is dimensionless, and the isothermal process is taken as 1; The insulation process is taken as 1.4.

### 3.3.2 *Motor Pump*

The isolation valve is used to block the energy connection between the brake master cylinder and the friction brake, and the motor pump is used as the electro-hydraulic brake energy supply device to provide hydraulic energy for the operation of the friction brake. The motor pump is composed of DC brushless motor and a high-speed piston pump. When establishing the mathematical model of motor pump dynamic characteristics, the influence of DC brushless motor dynamic characteristics is ignored, while the dynamic characteristics of the high-speed piston pump are expressed by the relationship between actual flow and actual pressure.

The pressure balance equation of the high-speed piston pump is:

$$P_P = R_P \Delta Q + L_P \frac{d\Delta Q}{dt} \tag{3.20}$$

where: $R_P$—hydraulic resistance of oil leakage circuit in hydraulic pump, N·s/m$^5$; $L_P$—Liquid sensing of oil leakage circuit in hydraulic pump, N·s$^2$/m$^5$; $\Delta Q$—Leakage of the pump, m$^3$/s.

The flow continuity equation of the high-speed piston pump is as follows:

$$Q_P = \frac{V_P}{K_P} \cdot \frac{dP_P}{dt} \tag{3.21}$$

where: $Q_P$—actual flow of the hydraulic pump, m$^3$/s; $V_P$—Volume of pressure oil chamber of hydraulic pump, m$^3$; $K_P$—Equivalent bulk modulus of elasticity of hydraulic pump, Pa.

### 3.3.3 Brake Master Cylinder

As Fig. 3.5 shows, the isolation valve is opened when the electro-hydraulic brake fails, and the pedal force applied by the driver on the brake pedal will be transmitted

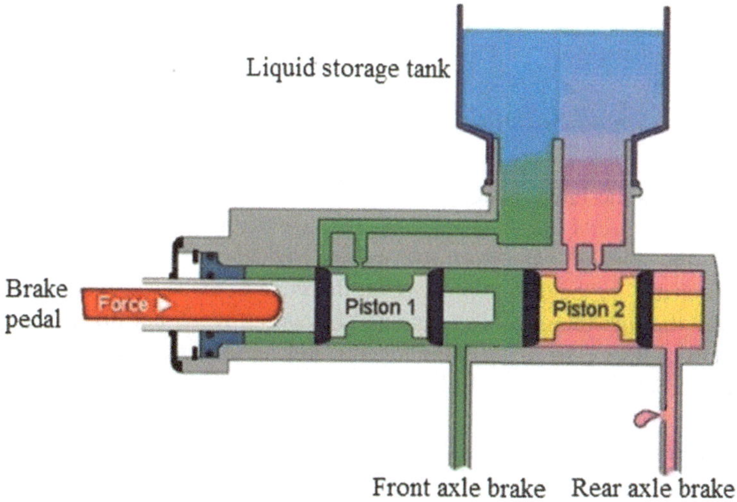


**Fig. 3.5** Schematic diagram of the master cylinder's structure

to piston 1 through the push rod of the brake master cylinder. Piston 1 under the pushrod force will continue to move and pressurize the brake fluid, and the high-pressure brake fluid will be transmitted to the front axle friction brake through the brake pipeline; At this time, the ejector rod of piston 1 does not move to eliminate the gap with piston 2, so there is no braking pressure in the rear axle friction brake. At this time, piston 1 and piston 2 can be regarded as two independent rigid bodies and analyzed.

The pressure balance equation of the brake master cylinder is as follows:

$$\begin{cases} m_{mr}\frac{d^2x_{mr}}{dt^2} = F_M - P_{mr}A_{mr} - F_{msr} - F_{mbr} \\ m_{mf}\frac{d^2x_{mf}}{dt^2} = P_{mf}A_{mf} + F_{msr} + F_{mbr} - P_{mf}A_{mf} - F_{msf} - F_{mbf} \end{cases} \tag{3.22}$$

where: $m_{mr}$—mass of rear cylinder piston, kg; $x_{mr}$—Displacement of rear cylinder piston, m; $F_M$—Input force of brake master cylinder pushrod, N; $P_{mr}$—Hydraulic pressure of rear cylinder, Pa; $A_{mr}$—Cross-sectional area of rear cylinder piston, m$^2$; $F_{msr}$—Spring force of return spring of rear cylinder, N; $F_{mbr}$—Damping force of rear cylinder piston, N; $m_{mf}$—Mass of front cylinder piston, kg; $x_{mf}$—Displacement of front cylinder piston, m; $F_{msf}$—Spring force of return spring of front cylinder, N; $F_{mbf}$—Damping force of front cylinder piston, N; $A_{mf}$—Cross-sectional area of front cylinder piston, m$^2$; $P_{2mf}$—hydraulic pressure of front cylinder, Pa.

The flow continuity equation of the brake master cylinder is:

$$\begin{cases} Q_{mf} = A_{mf}\frac{dx_{mf}}{dt} \\ Q_{mr} = A_{mr}\frac{dx_{mr}}{dt} \end{cases} \tag{3.23}$$

where: $Q_{mf}$—actual flow of front cylinder, m$^3$/s; $Q_{mr}$—Actual flow of rear cylinder, m$^3$/s.

### 3.3.4 *Low-Speed Solenoid Valve*

In electro-hydraulic brake, since the isolation valve and balance valve only start to work under special circumstances, and the switching frequency is low during operation, the solenoid valve with low natural frequency (i.e., low-speed solenoid valve) is selected in the design process. At the same time, because the flow rate of brake fluid through the isolation valve and balance valve is not high, they can be regarded as orifices. The flow continuity equations of the low-speed solenoid valve are given according to the state of brake fluid passing through orifices.

When the flow of brake fluid at the orifice is turbulent, the flow continuity equation of low-speed solenoid valve is:

$$Q_V = C_d A_V \sqrt{\frac{2\Delta p}{\rho_f}} \tag{3.24}$$

where: $Q_V$—orifice flow, $m^3/s$; $C_d$—Flow coefficient; $A_V$—Sectional area of orifice, $m^2$; $\Delta p$—Pressure difference at both ends of the orifice, Pa; $\rho_f$—Brake fluid density, $kg/m^3$.

When the flow of brake fluid at the orifice is laminar flow, the flow continuity equation of low-speed solenoid valve is:

$$Q_V = \frac{8C_d^2 A_V^2 \Delta p}{v_f \rho_f \chi R_{ef}} \tag{3.25}$$

where: $R_{ef}$—critical Reynolds number of brake fluid; $v_f$—Kinematic viscosity of brake fluid, $m^2/s$; $\chi$—Wet circumference length of brake pipeline, m;

### 3.3.5 High-Speed Solenoid Valve

The high-speed solenoid valve is the core component to realize the pressure regulation of the brake wheel cylinder in electro-hydraulic brake. As Fig. 3.6 [1] shows, the high-speed battery valve has only two working states: open and closed. In the open state, the oil outlet of the high-speed solenoid valve is connected with the oil drain port, the brake fluid flows back to the brake fluid tank through the oil drain port, and in the closed state, the oil outlet of the high-speed solenoid valve is connected

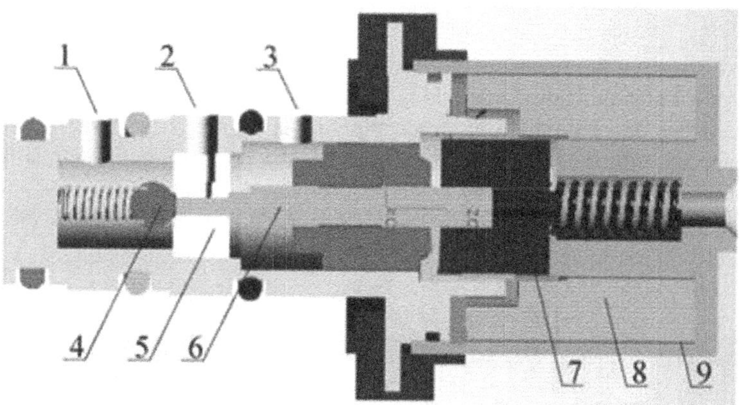


1. Oil inlet 2. Oil outlet 3. Oil drain 4. Valve ball 5. Valve seat 6. Valve stem 7. Armature 8. Coil 9. Magnetic conducting shell

**Fig. 3.6** Schematic diagram of high-speed switch valve's structure

with the oil inlet. The brake enters the brake wheel cylinder through the oil outlet. The high-speed solenoid valve accurately controls the brake fluid flow through the high-speed solenoid valve by quickly opening and closing to achieve the effect of accurate control of wheel motion. The continuous flow equation of a high-speed solenoid valve is:

$$\begin{cases} Q_{vi} = C_d A_{vi} \sqrt{\frac{2(P_{vi} - P_{val})}{\rho_f}} \operatorname{sgn}(P_{val} - P_{vo}) \\ Q_{vo} = C_d A_{vo} \sqrt{\frac{2(P_{val} - P_{vo})}{\rho_f}} \operatorname{sgn}(P_{vi} - P_{val}) \end{cases} \tag{3.26}$$

where: $Q_{vi}$, $Q_{vo}$—actual flow at oil inlet and outlet of the high-speed solenoid valve, $m^3/s$; $A_{vi}$, $A_{vo}$—Sectional area of oil inlet and outlet of the high-speed solenoid valve, $m^2$; $P_{vi}$, $P_{vo}$—Actual pressure of oil inlet and outlet of the high-speed solenoid valve, Pa; $P_{val}$—Actual pressure of hydraulic chamber of the high-speed solenoid valve, Pa;

If the compressibility of the brake fluid causes the oil pressure in the hydraulic chamber to change, the pressure balance equation of the high-speed solenoid valve is:

$$P_V = \frac{K_f}{V_V} \int (Q_{vi} - Q_{vo}) dt + P_{val} \tag{3.27}$$

where: $V_V$—volume of the hydraulic chamber of the high-speed solenoid valve, $m^3$; $K_f$—Bulk elastic modulus of brake fluid, Pa.

### 3.3.6 Brake Line

When investigating the pressure and flow characteristics of the brake pipeline, the centralized parameter analysis method is used to simplify the modeling of the brake pipeline. If the brake fluid in the brake pipeline is regarded as a whole without considering the influence of brake fluid viscosity, the pressure and flow rate in the whole brake pipeline is equal everywhere.

The pressure balance equation and fluid continuity equation of the brake pipeline are respectively expressed as:

$$\begin{cases} P_{hi} = P_{ho} + R_H Q_{ho} + L_H \frac{dQ_{ho}}{dt} \\ Q_{hi} = Q_{ho} + C_H \frac{dP_{ho}}{dt} \end{cases} \tag{3.28}$$

where: $P_{hi}$, $P_{ho}$—Inlet and outlet pressure of brake pipeline, Pa; $R_H$—Hydraulic resistance of brake pipeline, $N \cdot s/m^5$; $L_H$—Hydraulic sensing of brake pipeline, $N \cdot s^2/m^5$; $Q_{hi}$, $Q_{ho}$—Inlet and outlet flow of brake pipeline, $m^3/s$; $C_H$—Brake pipeline fluid capacity, $m^3/N$.

### 3.3.7 Brake Wheel Cylinder

The brake pressurization process of the brake wheel cylinder can be divided into two stages. The first is to eliminate the empty stroke of the brake clearance between the brake lining and the brake disc, and then the stable pressurization stage in which the brake lining and the brake disc are deformed. As shown in Fig. 3.7 [2], in the idle stroke stage, the piston of the brake wheel cylinder first overcomes the friction and starts to move. At this time, the pressure from the accumulator is only used to balance various resistances during the movement of the piston until the brake clearance disappears. Regardless of the compressibility of the brake fluid, the flow equation of the brake wheel cylinder in the idle stroke stage is:

$$Q_C = A_C \frac{dX_C}{dt} \tag{3.29}$$

where: $A_C$, $X_C$—cross-sectional area and displacement of brake wheel cylinder piston, m; $Q_C$—Flow of brake wheel cylinder, m$^3$/s.

For disc brake, the force and movement of the piston in wheel cylinder can be approximately indicated by mass spring damper system. Assuming that the displacement of piston is zero at the beginning of piston movement, the pressure balance equation of brake wheel cylinder can be expressed as:

$$P_C A_C = m_C \frac{d^2 X_C}{dt^2} + B_C \frac{dX_C}{dt} + (K_d + K_s) X_C \tag{3.30}$$

Fig. 3.7 Schematic diagram of disc brake's structure

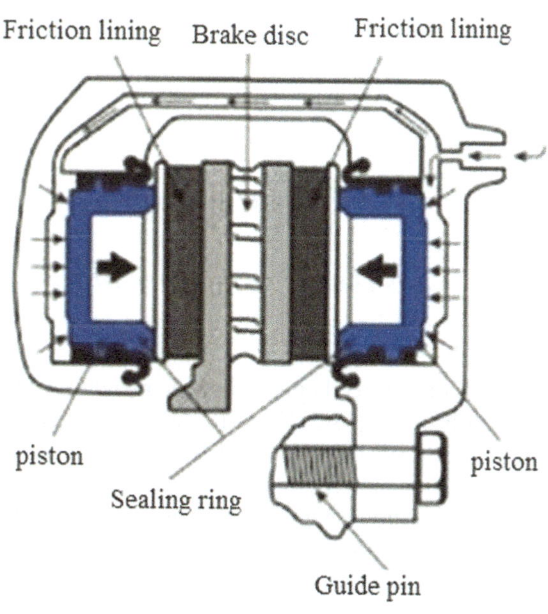

where: $P_C$—pressure of brake wheel cylinder, Pa; $m_C$—Mass of piston, kg; $B_C$—Motion damping coefficient of the piston; $K_d$—Equivalent spring stiffness of brake disc, N/m; $K_s$—Stiffness of return spring, N/m.

At the end of the idle stroke, the motion output of the piston of the brake wheel cylinder is very small. When analyzing the pressure and flow characteristics of the wheel cylinder, the compressibility of brake fluid and the elastic deformation of the brake itself become the main factors of the wheel cylinder dynamic mechanics. Supposing that the brake fluid inflow direction of the brake wheel cylinder is positive, the flow continuity equation of the brake wheel cylinder is:

$$Q_C = \frac{V_C}{K_C}\frac{dP_C}{dt} \tag{3.31}$$

where: $V_C$—brake wheel cylinder volume, m$^3$; $K_C$—Equivalent bulk elastic modulus of the brake wheel cylinder, Pa.

Taking the equivalent spring stiffness of the brake into consideration, the calculation formula of the equivalent bulk elastic modulus of the brake wheel cylinder is:

$$\frac{1}{K_C} = \frac{1}{K_f} + \frac{A_C^2}{K_d V_C} \tag{3.32}$$

Be based on the pressure of the brake wheel cylinder, the output braking torque of the disc brake can be calculated as:

$$T_h = 2\eta_C R_d P_C A_C \tag{3.33}$$

where: $\eta_C$—friction coefficient between the brake lining and brake disc; $R_d$—Effective radius of the brake disc, m.

## 3.4 Electro-Pneumatic Brake Model

### *3.4.1 Series Double—Chamber Brake Valve*

The structure of the series dual-chamber brake valve is shown in Fig. 3.8 [3]. The upper piston is directly pushed by the mandrel, while the small piston is pushed by the high-pressure gas on the upper side of the piston and the upper valve core.

Before establishing the simulation model of series double-cavity brake valve, the following assumptions are made:

(1) The gas flowing through the valve port or other flow limiting holes is isentropic flow; (2) The pressure distribution and temperature field of the gas in the chamber are uniform; (3) The friction and deformation resistance of the diaphragm is relatively

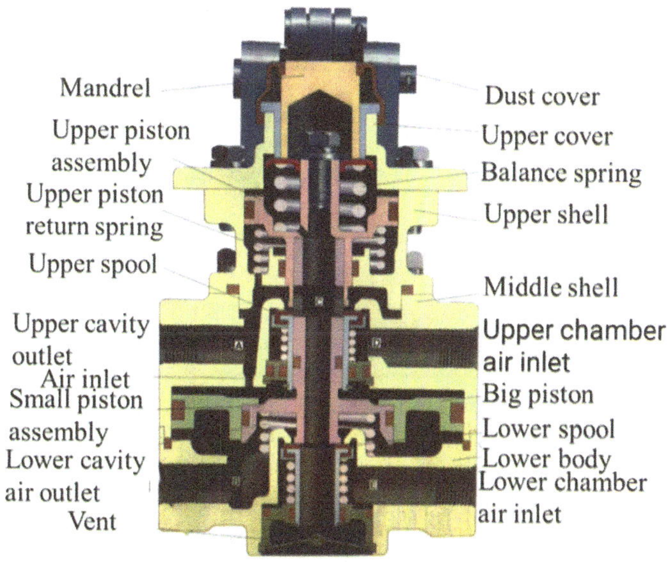


**Fig. 3.8** Schematic diagram of dual brake valve

small and can be ignored; (4) good sealing between the inlet and outlet valve and the rubber spool; (5) the weight of moving parts in the valve is very small relative to the force and can be ignored.

First of all, the motion force of the internal parts of the brake valve is analyzed. Here, the push rod, piston and valve group are regarded as a whole, and the equation can be listed as follows:

$$m\frac{d^2x}{d^2t} + B\frac{dx}{dt} + F_1 + F_2 + F_3 + k_1x_1 + k_2x_2 + k_3x_3 + P_AS_A + P_BS_B = F \quad (3.34)$$

where: $m$—the total mass of M-piston components, kg; $B$—damping coefficient; $F_1$, $F_2$ and $F_3$—the preloading forces of the return springs of the upper chamber piston, the lower chamber piston and the exhaust valve, N; $k_1$, $k_2$ and $k_3$—the stiffness of the return springs, N/m; $P_A$ and $P_B$—output pressures of upper and lower cavities, Pa; $S_A$ and $S_B$—the action area of the output pressure of the upper and lower cavities, m$^2$; $F$—push rod force, N, $F = kx_{push} = k\,(x_0 + x_1)$, $k$—the rubber spring stiffness, N/m; $x_{push}$—the push rod displacement, m; $x_0$—the rubber spring precompression, m, $x$—the piston assembly displacement, m; $x_1$, $x_2$, $x_3$—the displacement of the upper chamber piston, the lower chamber piston and the return spring of the exhaust valve, m.

Secondly, the mass flow rate of each valve port gas is analyzed, and the mass flow rate of the upper valve port (the lower valve port is the same as above) is obtained according to the mass flow formula of the gas isentropic passing through the shrinking nozzle:

$$\begin{cases} Q_A = \mu A(x_A)P_D\sqrt{\frac{2K}{RT(K-1)}\left[\left[\frac{P_A}{P_D}\right]^{\frac{2}{K}} - \left[\frac{P_A}{P_D}\right]^{\frac{X+1}{K}}\right]}, 1 > \frac{P_A}{P_D} > 0.528 \\ Q_A = \mu A(x_A)P_D\sqrt{\frac{2K}{RT(K+1)}\left(\frac{2}{K+1}\right)^{\frac{1}{K-1}}}, \frac{P_A}{P_D} \le 0.528 \end{cases} \quad (3.35)$$

Among them:

$$\begin{cases} A(x_A) = \pi\left(d_1^2 - d_2^2\right)/4, \pi d_2 x_A > \pi\left(d_1^2 - d_2^2\right)/4 \\ A(x_A) = \pi d_3 x_A, \pi d_3 x_A \le \pi\left(d_1^2 - d_2^2\right)/4 \end{cases} \quad (3.36)$$

where: $Q_A$—the atmospheric mass flow rate of the upper cavity, kg/s; $x_A$—the opening of the upper valve, m; $P_D$ and $P_A$—the pressure at the inlet and outlet of the valve port, Pa; $d_1$—spool outer diameter, m; $d_2$—the inner diameter of the spool, m; $d_3$—the inlet valve diameter, m; $\mu$—flow coefficient; $K$—the isentropic index, which is 1.4. $T$—the temperature of the gas, K; $R$—the gas constant, and for air $R$ is 287 J/(kg K).

Finally, the change of gas pressure in the upper and lower chambers of the brake valve with time is analyzed. Since the process time is short, the change of gas temperature is ignored here. Therefore, the relationship between the gas pressure in the upper chamber of the brake valve and time can be expressed as follows:

$$\frac{dF_A}{dt} = P_D\left[\frac{Q_A}{Q_{AO} + \int Q_A dt} + \frac{S_A}{V_{AO} - S_A x_A} \cdot \frac{dx_A}{dt}\right] \quad (3.37)$$

where: $V_{AO}$—the initial volume of the upper cavity, m$^3$; $Q_{AO}$—the initial gas mass flow rate, kg/s; $Q_{AO} = V_{AO}P_{AO}/RT$,where $P_{AO}$—the initial pressure of the upper cavity, Pa.

### 3.4.2 Relay Valve

Relay valve structure, as shown in Fig. 3.9 [4]. The relay valve is a control valve connected between the brake valve and the brake chamber. Its main function is to increase the response speed of the brake chamber pressure.

Assume a tandem double—chamber brake valve. The control pressure input from the 4 port is $P_4$, the pressure input from the 1 port gas storage tank is $P_1$, and the brake pressure output from the 2 port is $P_2$. The balance equation of the relay valve is as follows:

$$P_4\frac{\pi}{4}D_1^2 + P_2\frac{\pi}{4}\left(d_2^2 - d_4^2\right) = P_2\frac{\pi}{4}\left(D_1^2 - d_4^2\right) + F_2 + E_1 + P_1\frac{\pi}{4}\left(d_2^2 - d_1^2\right) \quad (3.38)$$

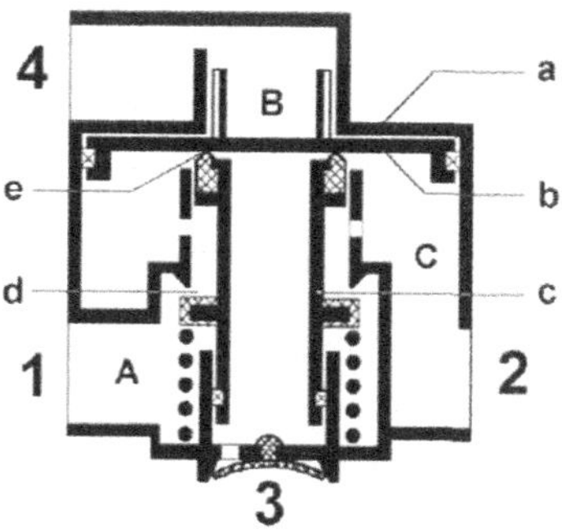


a. The cover b. Servo piston c. The valve core assembly d. Intake valve e. The exhaust valve 1. Answer the gasholder 2. Connecting brake line 3. Through the atmosphere 4. Connect series double - chamber brake valve

**Fig. 3.9** Relay valve

Thus, the working pressure required by the relay quick release valve at the beginning of work can be calculated. At this time, the air inlet has not started, so $P_2$ is zero, and the piston moving spring force needs to be overcome, so the expression is as follows:

$$P_{4\min}\frac{\pi}{4}D_1^2 - F_2 = P_1\frac{\pi}{4}\left(d_2^2 - d_1^2\right) + F_1 \tag{3.39}$$

where: $D_1$—the piston diameter, m; $D_3$—the inner diameter of the inlet valve door, m; $D_4$—the outer diameter of core body, m; $D_2$—inlet valve door outer diameter, m; $D_1$—the outer diameter of the pressure reducing rod, m; $F_2$—the piston reversing spring force, N; $F_1$—valve reversing spring force, N.

## 3.4.3 *Brake Chamber*

Electro-Pneumatic Brake usually adopts single diaphragm brake chamber and spring diaphragm brake chamber. The structures of the single-diaphragm brake chamber and spring diaphragm brake chamber are respectively shown in Figs. 3.10 [3] and 3.11 [3].

Dynamic thrust force $F_t$ and stroke of brake chamber can be fitted into a piecewise function:

$$F_t = \begin{cases} k_1 x_\text{t} & (x_\text{t} \le x_\text{tp}) \\ (k_1 - k_2)x_\text{tp} + k_2 x_\text{t} & (x_\text{t} > x_\text{tp}) \end{cases} \tag{3.40}$$

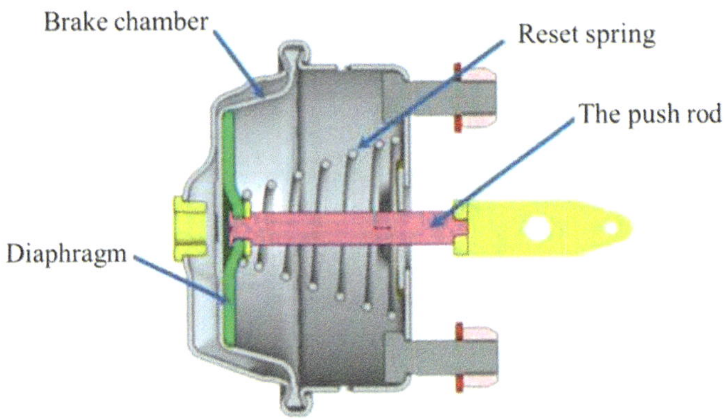


**Fig. 3.10** Schematic diagram of brake chamber

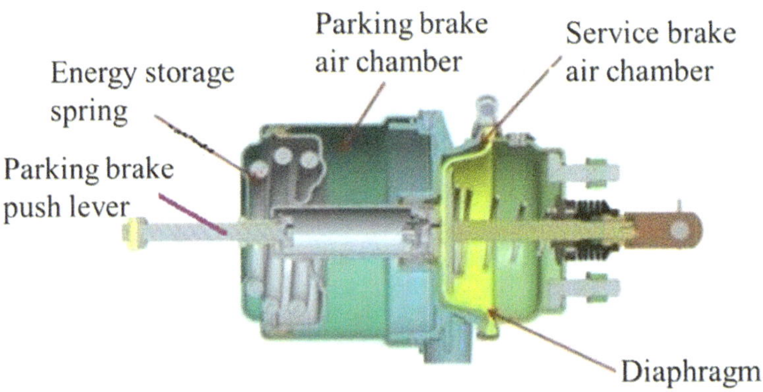


**Fig. 3.11** Schematic diagram of spring brake actuator

where: $k1$—the stiffness in the low stiffness region, which is related to the reset spring stiffness of the brake air chamber and the reset spring stiffness of the brake, N/mm; $k2$—the stiffness in the high stiffness region, which is related to the friction plate, brake shoe, brake drum elastic modulus and the motion transmission ratio of the CAM actuating mechanism, N/mm; $x$t—the push rod stroke, mm; $x$TP—a push rod stroke, used to eliminate the gap between the auto-adjusting arm and the hoof drum, mm.

### *3.4.4 Drum Brake*

The calculation formula of brake torque of drum brake is as follows:

$$T_b = 2K_{b2}T_{bt}R_b\eta_b/[y_b(1+i)] \tag{3.41}$$

where: $T_b$—the brake torque, Nm; $i = K_{b1}/K_{b2}$, where $K_{b1}$ and $K_{b1}$ are braking efficiency factors of lead shoe and slave shoe respectively. $R_b$—the brake drum radius, m; $y_b$—the effective radius of CAM, m; $\eta_b$—the braking efficiency coefficient.

In Eq. (3.41), the calculation formula of brake camshaft torque $M_{bt}$ is as follows:

$$M_{bt} = F_t \cdot L_b \tag{3.42}$$

where: $F_t$—the thrust of the brake air chamber push rod, N; $L_b$—the length of the brake arm, m.

## 3.5 Vehicle Brake Dynamic Model

### *3.5.1 Vehicle Longitudinal Dynamic Model*

When a fuel cell car moves, it is subjected to rolling resistance, acceleration resistance, and air resistance, and when the car is driving on a tilted road, its motion status is affected by gravity extension of the road direction. During the driving time of a fuel cell vehicle, the driving force provided by the motor is equal to the resistance of the vehicle and can be expressed in the formula (3.43)

$$F_{mot} = F_{air} + F_f + F_s + F_a. \tag{3.43}$$

where: $F_{mot}$—motor drive, $F_{air}$—air resistance, $F_f$—rolling resistance, $F_s$—slope resistance, $F_a$—acceleration resistance.

According to the principle of force balancing when the car is driving, its driving force is equal to the vector and all resistance. The motor supplies the driving force of the car, so its power can be expressed in type (3.44):

$$P_m = \frac{F_{mot} \cdot v}{n \cdot \eta_{mot}}. \tag{3.44}$$

where: $P_m$ represents the demand power that continues when the car is driving, $n$ for the drive ratio of the gearbox and main gearbox, and $\eta_{mot}$ for motor drive efficiency.

Motor demand power is provided by power cells and fuel cells and can be expressed in type (3.45):

$$P_{mot} = P_{bat} + P_{fc} \cdot \eta_{DCDC}. \tag{3.45}$$

where: $P_{fc}$ represents the power needed for the fuel cell, $P_{bat}$ represents the power needed for the battery, and $\eta_{DCDC}$ represents the energy transfer efficiency of the DCDC converter.

### 3.5.2 Vehicle Single-Wheel Braking Model

As shown in Fig. 3.12 [5], taking integrated brake system for electromagnetic and friction as an example, so as to predigest the analysis of the vehicle braking process, the yaw dynamics and pitch dynamics of the vehicle are not investigated, and the longitudinal dynamics of the vehicle are mainly considered. Therefore, the single wheel braking model is adopted in this paper, and the state equation during vehicle braking can be expressed as:

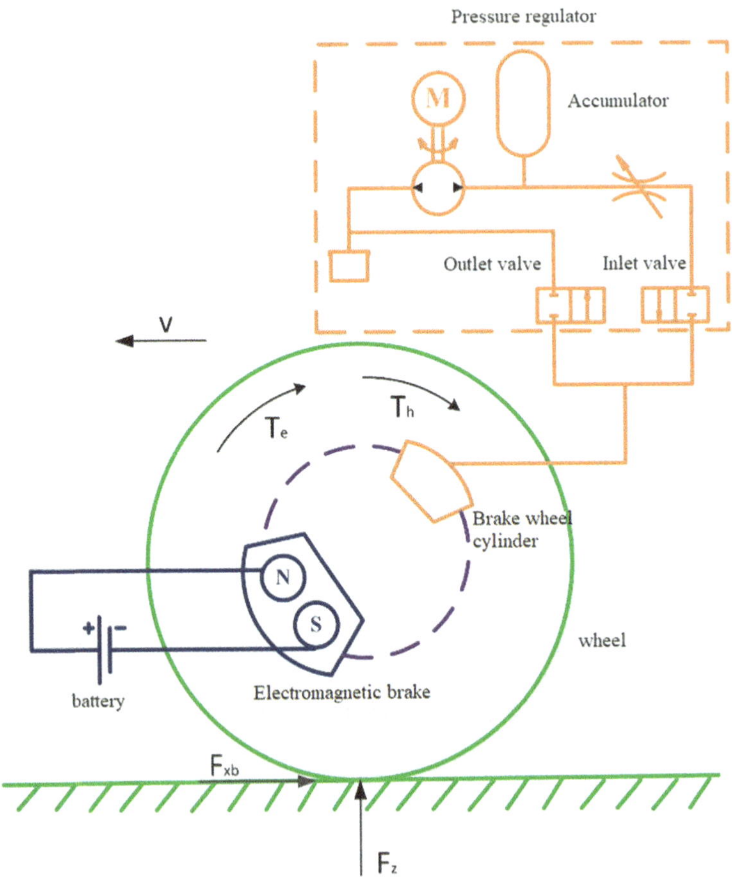


**Fig. 3.12** Single wheel brake model

$$\begin{cases} \dot{\omega} = \frac{1}{J_W}\left[F_z \cdot \varphi(\lambda) \cdot R_{roll} - T_a - T_h\right] \\ \dot{v} = -\frac{1}{M_V}\left[F_z \cdot \varphi(\lambda) + F_f + F_w\right] \end{cases} \tag{3.46}$$

where: $\omega$—wheel rotation angular velocity, rad/s; $J_W$—Wheel moment of inertia, kg m$^2$; $F_z$—Vertical force between tire and ground, N; $\varphi(\lambda)$—The adhesion coefficient is a function of the longitudinal slip ratio; $R_{roll}$—Tire rolling radius, m; $T_a$—Other forms of braking torque applied to the wheel, N m; $T_h$—Electro-hydraulic braking torque applied to the wheel, N m; $v$—Vehicle speed, m/s; $M_V$—Load mass of the single wheel, kg; $F_f$—Vehicle rolling resistance, N; $F_w$—Windward resistance, N.

The function of the adhesion coefficient varying with the longitudinal slip rate is generally characterized by the relationship between the braking force coefficient and the longitudinal slip rate. In the process of vehicle braking, the wheel motion state can be divided into three types: pure rolling, rolling while sliding and pure sliding. When the wheel is in the pure rolling state, the speed of the vehicle is equal to the speed of the wheels; when the vehicle is in the rolling and sliding state, there will be a certain difference between the vehicle speed and the wheel speed; when the wheel is in the pure sliding state, the wheel speed is zero, and the wheel is locked. Therefore, the proportion of sliding component in wheel motion is defined as the longitudinal slip rate, and the calculation expression of wheel longitudinal slip rate is as follows:

$$\lambda = \frac{v - R_{roll}\omega}{v} \tag{3.47}$$

where: $\lambda$—longitudinal slip ratio.

Meanwhile, the ratio of ground braking force to vertical load is defined as the braking force coefficient. As Fig. 3.13 [5] shows, the function of adhesion coefficient changing with longitudinal slip rate is generally simulated by the bilinear model.

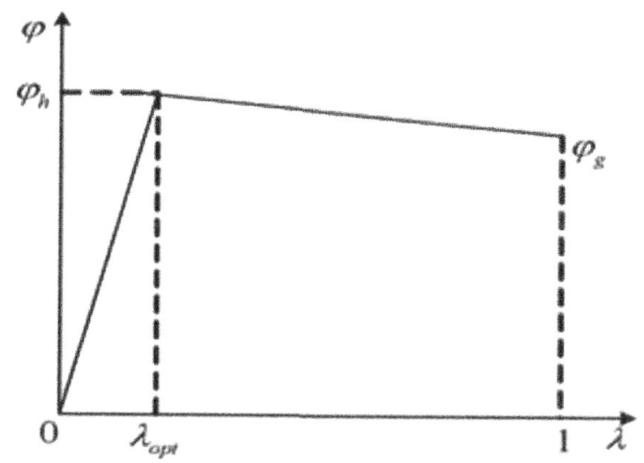


**Fig. 3.13** Simplified curve of the braking force coefficient and sliding rate

When the longitudinal slip rate is less than the optimal slip rate, the braking force coefficient increases with the increase of longitudinal slip rate, and the vehicle braking is in a stable area; when the longitudinal slip rate is larger than the optimal slip rate, the braking force coefficient decreases as the longitudinal slip rate increases, and the vehicle braking is in an unstable region.

The expression of the function of the adhesion coefficient varying with the longitudinal slip rate is:

$$\begin{cases} \varphi(\lambda) = \frac{\varphi_h}{\lambda_{opt}}\lambda & \lambda \le \lambda_{opt} \\ \varphi(\lambda) = \frac{\varphi_h - \varphi_g \lambda_{opt}}{1-\lambda_{opt}} - \frac{\varphi_h - \varphi_g}{1-\lambda_{opt}}\lambda & \lambda > \lambda_{opt} \end{cases} \tag{3.48}$$

where: $\varphi_h$—peak adhesion coefficient; $\lambda_{opt}$—optimal slip ratio; $\varphi_g$—sliding adhesion coefficient.

The simplified deformation of Eq. (3.48) is obtained:

$$\begin{cases} \varphi(\lambda) = \frac{\varphi_h}{\lambda_{opt}}\lambda \quad \lambda \le \lambda_{opt} \\ \varphi(\lambda) = \varphi_m - K_m\lambda \; \lambda > \lambda_{opt} \end{cases} \tag{3.49}$$

Of which: $\varphi_m = \frac{\varphi_h - \varphi_g \lambda_{opt}}{1-\lambda_{opt}}$

$$K_m = \frac{\varphi_h - \varphi_g}{1-\lambda_{opt}}$$

## 3.6 Hybrid Brake-By-Wire System Model

### *3.6.1 Model of Composite Brake System for Motor and Friction*

Considering the coupling nonlinear characteristics of composite brake system for motor and friction, the power bond graph (PBG) method is used. The coupling nonlinear model is established. The power bonding diagram of composite brake system for motor and friction is shown in Fig. 3.14 [6], while the power bonding diagram of electro-hydraulic brake under decompression and pressure holding is similar, which will not be repeated here.

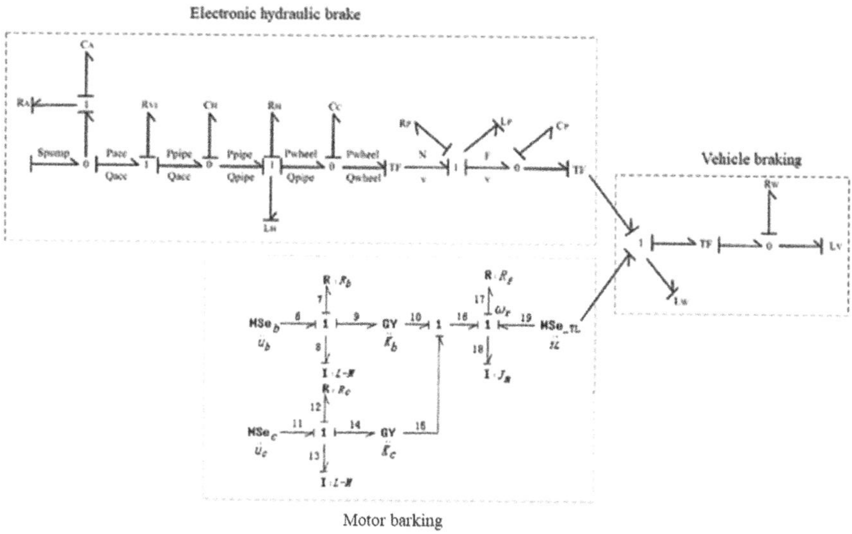


**Fig. 3.14** Power bond graph of composite brake system for motor and friction

The mathematical models of composite brake system for motor and friction under friction braking mode, motor braking mode and combined braking mode are obtained by using power bond graph method as follows.

(1) Friction braking mode

$$\begin{cases} \dot{V}_A = \frac{V_A}{C_A(R_V+R_A)} - \frac{V_H}{C_A(R_V+R_A)} \\ \dot{V}_H = \frac{V_A}{C_A(R_V+R_A)} - \frac{V_H}{C_A(R_V+R_A)} - \frac{P_H}{L_H} \\ \dot{P}_H = \frac{V_H}{C_H} - \frac{V_C}{C_C} - R_H\frac{P_H}{L_H} \\ \dot{V}_C = \frac{P_H}{L_H} - v_V A_V \\ \dot{v}_C = \frac{1}{M_C}\left[\frac{V_C}{C_C}A_C - B_C v_C - (K_d + K_s)X_C\right] \\ \dot{X}_C = v_C \\ \dot{\omega} = \frac{1}{J_W}\left[F_z \cdot \mu(\lambda) \cdot R_{roll} - 2\eta_C R_d \frac{V_C}{C_C} A_C\right] \\ \dot{v} = -\frac{1}{M_V}\left[F_z \cdot \mu(\lambda) + F_f + F_w\right] \end{cases} \tag{3.50}$$

(2) Motor braking mode

$$\begin{cases} \dot{\omega} = \frac{1}{J_W}\left[F_z \cdot \mu(\lambda) \cdot R_{roll} - k_{mo} \cdot \omega \cdot \phi^2\right] \\ \dot{v} = -\frac{1}{M_V}\left[F_z \cdot \mu(\lambda) + F_f + F_W\right] \end{cases} \tag{3.51}$$

(3) Combined braking mode

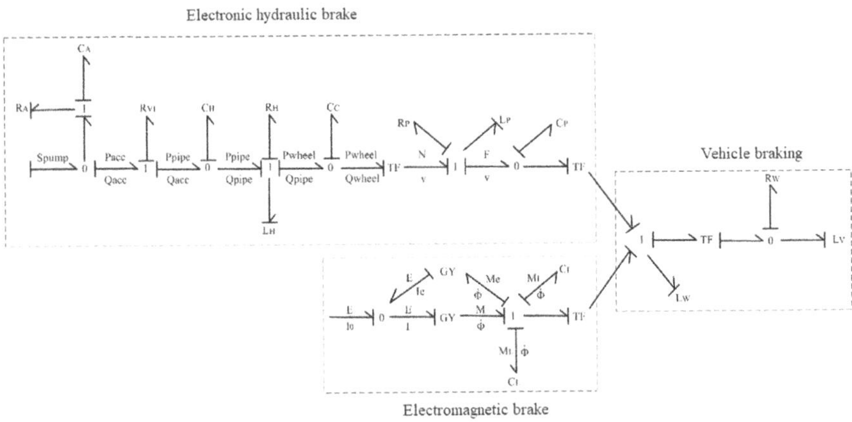


**Fig. 3.15** Power bond graph of integrated brake system for electromagnetic and friction

$$\begin{cases} \dot{V}_A = \frac{V_A}{C_A(R_V+R_A)} - \frac{V_H}{C_A(R_V+R_A)} \\ \dot{V}_H = \frac{V_A}{C_A(R_V+R_A)} - \frac{V_H}{C_A(R_V+R_A)} - \frac{P_H}{L_H} \\ \dot{P}_H = \frac{V_H}{C_H} - \frac{V_C}{C_C} - R_H \frac{P_H}{L_H} \\ \dot{V}_C = \frac{P_H}{L_H} - v_V A_V \\ \dot{v}_C = \frac{1}{M_C}\left[\frac{V_C}{C_C}A_C - B_C v_C - (K_d + K_s)X_C\right] \\ \dot{X}_C = v_C \\ \dot{\omega} = \frac{1}{J_W}\left[F_z \cdot \mu(\lambda) \cdot R_{roll} - 2\eta_C R_d \frac{V_C}{C_C} A_C - k_{mo} \cdot \omega \cdot \phi^2\right] \\ \dot{v} = -\frac{1}{M_V}\left[F_z \cdot \mu(\lambda) + F_f + F_W\right] \end{cases} \tag{3.52}$$

where: $V_A$—the gas volume of the accumulator, m$^3$; $V_H$—the liquid volume of brake pipeline, m$^3$; $R_V$—equivalent hydraulic resistance of high-speed solenoid valve, N s/m$^5$; $P_H$—liquid momentum of brake pipeline, kg m/s; $C_C$—liquid capacity of the brake wheel cylinder, m$^3$/N; $v_C$—movement speed of brake piston, m/s, $k_{mo}$ is the electric machine gain coefficient.

## 3.6.2 *Model of Integrated Brake System for Electromagnetic and Friction*

The coupled nonlinear model is established by power bond graph (PBG). The power bond graph of integrated brake system for electromagnetic and friction is shown in Fig. 3.15.

The mathematical models of integrated brake system for electromagnetic and friction under friction braking mode, electromagnetic braking mode and combined braking mode are as follows.

(1) Friction braking mode

$$\begin{cases} \dot{V}_A = \frac{V_A}{C_A(R_V+R_A)} - \frac{V_H}{C_A(R_V+R_A)} \\ \dot{V}_H = \frac{V_A}{C_A(R_V+R_A)} - \frac{V_H}{C_A(R_V+R_A)} - \frac{P_H}{L_H} \\ \dot{P}_H = \frac{V_H}{C_H} - \frac{V_C}{C_C} - R_H\frac{P_H}{L_H} \\ \dot{V}_C = \frac{P_H}{L_H} - v_V A_V \\ \dot{v}_C = \frac{1}{M_C}\left[\frac{V_C}{C_C}A_C - B_C v_C - (K_d + K_s)X_C\right] \\ \dot{X}_C = v_C \\ \dot{\omega} = \frac{1}{J_W}\left[F_z \cdot \mu(\lambda) \cdot R_{roll} - 2\eta_C R_d \frac{V_C}{C_C} A_C\right] \\ \dot{v} = -\frac{1}{M_V}\left[F_z \cdot \mu(\lambda) + F_f + F_w\right] \end{cases} \tag{3.53}$$

(2) Electromagnetic braking mode

$$\begin{cases} \dot{\omega} = \frac{1}{J_W}\left[F_z \cdot \mu(\lambda) \cdot R_{roll} - k_{eb} \cdot \omega \cdot \phi^2\right] \\ \dot{v} = -\frac{1}{M_V}\left[F_z \cdot \mu(\lambda) + F_f + F_W\right] \\ \dot{\phi} = \frac{\sqrt{2}}{2}B \cdot S_p \cdot \omega \end{cases} \tag{3.54}$$

(3) Combined braking mode

$$\begin{cases} \dot{V}_A = \frac{V_A}{C_A(R_V+R_A)} - \frac{V_H}{C_A(R_V+R_A)} \\ \dot{V}_H = \frac{V_A}{C_A(R_V+R_A)} - \frac{V_H}{C_A(R_V+R_A)} - \frac{P_H}{L_H} \\ \dot{P}_H = \frac{V_H}{C_H} - \frac{V_C}{C_C} - R_H\frac{P_H}{L_H} \\ \dot{V}_C = \frac{P_H}{L_H} - v_V A_V \\ \dot{v}_C = \frac{1}{M_C}\left[\frac{V_C}{C_C}A_C - B_C v_C - (K_d + K_s)X_C\right] \\ \dot{X}_C = v_C \\ \dot{\omega} = \frac{1}{J_W}\left[F_z \cdot \mu(\lambda) \cdot R_{roll} - 2\eta_C R_d \frac{V_C}{C_C} A_C - k_{eb} \cdot \omega \cdot \phi^2\right] \\ \dot{v} = -\frac{1}{M_V}\left[F_z \cdot \mu(\lambda) + F_f + F_W\right] \\ \dot{\phi} = \frac{\sqrt{2}}{2}B \cdot S_p \cdot \omega \end{cases} \tag{3.55}$$

where: $V_A$—the gas volume of the accumulator, m$^3$; $V_H$—the liquid volume of brake pipeline, m$^3$; $R_V$—equivalent hydraulic resistance of high-speed solenoid valve, N s/m$^5$; $P_H$—liquid momentum of brake pipeline, kg m/s; $C_C$—liquid capacity of the brake wheel cylinder, m$^3$/N; $v_C$—movement speed of brake piston, m/s, $k_{eb}$ is the electric machine gain coefficient.

## References

1. Minmin Z, Li C, Yi H et al (2010) Simplification and verification of nonlinear model of high-speed solenoid on-off valve. J Shanghai Jiaotong Univ (Chin Ed) 07:1005–1009
2. Yonghui Z (2010) Research on ABS matching process and calibration technology. Tsinghua University
3. Shanshan C (2017) Study on braking delay characteristics of heavy equipment transport vehicle combined braking system. Jiangsu University
4. Xianxiao H (2014) Modeling and simulation of gas-hydraulic hybrid braking system for multi-axle vehicle. Wuhan University of Technology
5. Jing H, Liu Z, Chen H (2011) A switched control strategy for antilock braking system with on/off valves. IEEE Trans Veh Technol 60(4):1470–1484
6. Karnopp DC, Margolis DL, Rosenberg RC (2012) System dynamics: modeling, simulation, and control of mechatronic systems. Wiley

# Chapter 4
# Design of Hybrid Brake-By-Wire System

The design of the control system of the intelligent vehicle hybrid line plays a vital role in creating intelligent vehicles. This chapter mainly introduces the safety design of composite brake system for motor and friction and the energy-saving design of integrated brake system for electromagnetic and friction.

## 4.1 Safety Design of Composite Brake System for Motor and Friction

### *4.1.1 Safety Design of Power Supply Device*

The energy supply device of the EHB system includes an airbag accumulator and a motor hydraulic pump. The design of the accumulator and motor pump is based on the requirements. Moreover, each can independently supply energy to the EHB under ordinary braking conditions in the event that the other fails, thus ensuring certain hardware redundancy. At the same time, it should be noted that the effective displacement of the accumulator and the rated flow of the hydraulic pump are mutually coupled. At the beginning of the design process, the braking performance requirements of the target vehicle should be fully considered. Only then can the maximum and minimum operating pressures for the EHB system be determined.

(1) Accumulator charging pressure: Determine the maximum and minimum pressure in the system. For the minimum working point pressure, the accumulator skins have yet to swell to contact the shell's inner wall. At the same time, the highest pressure point is skin contraction after a quarter of the volume is still more significant than the initial volume. The scope of the accumulator charging pressure is the following:



D. Hu, *Design and Control of Hybrid Brake-by-Wire System for Autonomous Vehicle*, Key Technologies on New Energy Vehicles,
https://doi.org/10.1007/978-981-16-8946-8_4

$$0.25\ P_1 < P_0 < 0.9\ P_2 \tag{4.1}$$

where: *P0* is the accumulator precharging pressure; *P1* is the highest working pressure of the system; and *P2* is the minimum working pressure of the system.

(2) Determination of the effective volume of the accumulator: It is assumed that the discharge process of the accumulator is fast, which is approximately an adiabatic process. Compared with gas, the compressibility of oil can be ignored; Boyle's law of thermodynamics gives

$$P_0 V_0^n = P_1 V_1^n = P_2 V_2^n \tag{4.2}$$

where: *V0* is the effective volume of the accumulator; *V1* is the minimum gas volume of the accumulator; and *V2* is the maximum gas volume of the accumulator. Substituting the effective displacement of the accumulator, $\Delta V = V_2 - V_1$, into Eq. (4.2), the calculation formula of the effective volume of the accumulator can be obtained as follows:

$$V_0 = \frac{\Delta V}{P_0^{0.715}\left[\left(\frac{1}{P_2}\right)^{0.715} - \left(\frac{1}{P_1}\right)^{0.715}\right]} \tag{4.3}$$

The effective displacement of the accumulator should meet the following two conditions. When the motor hydraulic pump fails, the number of times that the accumulator can complete high-strength braking should not be less than 4–5 times. When the motor hydraulic pump is working, after 12 brakings at a rate of 4 times per minute, the pressure supplied to the brake shall not be less than 70% of the initial measured pressure. Considering that the brake wears gradually in actual use, the calculation is carried out with the old brake after complete wear to ensure safety.

The calculation formula of the effective displacement of the accumulator is as follows:

$$\Delta V = \sum_{i=1}^{n} V_i - \sum_{j=1}^{n} Q_p t_p \tag{4.4}$$

where: *Vi* is the volume of brake fluid required during braking and *tp* is the working time of the hydraulic pump.

(3) Determination of the rated flow of the hydraulic pump: The filling time of the motor hydraulic pump should be reasonably determined. Generally, the filling time should not exceed 5 s. At the same time, it should be noted that in the case of accumulator failure. The motor pump should be able to implement the braking requirements of ordinary braking conditions. After the design's completion, the rated flow of the hydraulic pump should meet the requirements. The calculation formula of the rated flow of the hydraulic pump is

$$Q_p = \varepsilon \frac{\Delta V}{t_p} \tag{4.5}$$

where: $\varepsilon$ is the leakage coefficient of the hydraulic pump;

(4) Determination of motor-rated power: The highest working pressure of the system is only the static working pressure of the system. The rated working pressure of the hydraulic pump should be approximately 1.25 times the highest working pressure of the system. The power of the motor hydraulic pump should be calculated as follows:

$$P_e = \frac{P_{pm} \times Q_p}{60\eta_p} \tag{4.6}$$

where: $\eta_p$ is the efficiency of the motor hydraulic pump and *Ppm* is the rated pressure of the motor hydraulic pump.

### 4.1.2 Safety Design of Spare Brake Circuits

By stepping on the brake pedal, the driver forces the brake fluid in the brake master cylinder into the brake wheel cylinder through the isolation valve and the brake pipeline. In the case of a power supply device failure, the backup braking circuit needs to support a certain braking capability. When the pedal force reaches 500 N, the braking deceleration speed of the vehicle should be greater than the legal requirement of 3 m/s$^2$.

(1) Determining the cross-sectional area of the brake master cylinder: When the transmission ratio of the brake pedal and the structural parameters of the wheel brake are determined, the cross-sectional area of the brake master cylinder directly affects the output braking torque of the standby brake circuit.

First, the calculation formula of the ground braking force generated by a single wheel is given as follows:

$$\mathrm{F_x} = \mathrm{P_C} \cdot \mathrm{A_C} \cdot \eta_\mathrm{C} \cdot \mathrm{C}^* \cdot \frac{\mathrm{r_m}}{\mathrm{R_{roll}}} \tag{4.7}$$

where: *Fx* is the ground braking force generated by the brake; $C^*$ is the brake gain coefficient; *Rroll* is the wheel rolling radius; $\eta_c$ is the mechanical efficiency of the brake; and *rm* is the effective radius of the brake disc.

To meet the legal requirements of a 500 N pedal force more significant than 3 m/s$^2$, the ground-braking force of the front and rear axles is required to meet the following criteria:

$$\mathrm{F_{xf}} + \mathrm{F_{xr}} = \mathrm{Gz} \tag{4.8}$$

where: $Fxf$ is the ground braking force of the front axle; $Fxr$ is the ground braking force of the rear axle; $z$ is the brake strength; and $G$ is vehicle gravity at full load.

Substituting Eq. (4.7) into Eq. (4.8), the output pressure of the brake master cylinder can be obtained:

$$P_m = \frac{MaR_{roll}}{2\eta_c C^*\left(A_{cf}r_{ef} + A_{cr}r_{er}\right)} \tag{4.9}$$

where: $Pm$ is the brake master cylinder pressure; $Acf$ and $Acr$ are the cross-sectional areas of the front and rear brake wheel cylinder pistons, respectively; and $ref$ and $rer$ are the effective friction radii of the front and rear brake discs, respectively.

Then, the cross-sectional area of the brake master cylinder can be estimated by the following formula:

$$A_m = \frac{F}{P_m} = \frac{F_{\text{ped}} \cdot i_{\text{ped}}}{P_m}\eta_{\text{ped}} \tag{4.10}$$

where: $Fped$ is the driver's control force; $iped$ is the brake pedal transmission ratio; and $\eta_{ped}$ is the mechanical efficiency of the brake pedal.

(2) Determining the cross-sectional area of the isolation valve: The rated pressure of the isolation valve is calculated according to Eq. (4.9). The rated flow shall meet the braking response time requirements of the system. The calculation formula of the rated flow of the isolation valve is as follows:

$$\begin{cases} Q_{\text{gf}} = \frac{2\text{V}_{\text{cf}}}{\text{t}_{\text{m}}} \\ \text{Q}_{\text{gr}} = \frac{2\text{V}_{\text{cr}}}{\text{t}_{\text{m}}} \end{cases} \tag{4.11}$$

where: $tm$ is the braking response time of the system; $Vcf$ and $Vcr$ are the amount of fluid required by the front and rear axle wheel brakes, respectively;

In combination with Eq. (4.4), the calculation formula for the cross-sectional area of the front and rear axle isolation valves is as follows:

$$\begin{cases} A_{gf} = \sqrt{\frac{\nu\rho\chi R_e Q_{gf}}{8C_d^2 P_m}} \\ A_{gr} = \sqrt{\frac{\nu\rho\chi R_e Q_{gr}}{8C_d^2 P_m}} \end{cases} \tag{4.12}$$

where: $Agf$ is the cross-sectional area of the front axle isolation valve and $Agr$ is the cross-sectional area of the rear axle isolation valve.

### 4.1.3 Safety Design of Inlet/Outlet Valve

For solenoid valves, the rated flow, maximum working pressure, and natural frequency are essential performance indicators. The maximum working pressure of the inlet valve and outlet valve depends on the maximum working pressure of the system. The selection of the solenoid valve's natural frequency is required to meet the requirements of anti-lock braking control. Formula (4.13) notes that the rated flow of a high-speed solenoid valve is determined by the pressure difference of single braking, that is, the pressurization rate of the inlet valve and the decompression rate of the outlet valve. In the case of emergency braking, it is generally required that the EHB be pressurized as fast as possible. However, in the actual design, the control lag phenomenon of the hydraulic system should also be considered. Excess pressurization speed will lead to the problem of pressure overshoot. Generally, the pressurization rate is required to be 75 MPa/s during anti-lock braking.

First, the calculation formula of the maximum pressure of the front and rear brake wheel cylinders under emergency braking is given according to the ideal braking force distribution relationship of the front and rear axle brakes.

$$\begin{cases} P_f = \dfrac{G\varphi(b+\varphi h_g)R_{roll}}{2(a+b)\cdot A_c\cdot \eta_c\cdot C^*\cdot r_m} \\ P_r = \dfrac{G\varphi(a-\varphi h_g)R_{roll}}{2(a+b)\cdot A_c\cdot \eta_c\cdot C^*\cdot r_m} \end{cases} \tag{4.13}$$

where: $Pf$ and $Pr$ are the brake wheel cylinder pressures of a single wheel on the front and rear axles, respectively; $\varphi$ is the road adhesion coefficient; $a$ and $b$ are distanced from the center of mass to the front and rear axes, respectively; and $hg$ is the height of the center of mass.

By combining Eqs. (4.3) and (4.22), the calculation formula for the cross-sectional area of the inlet valve is as follows:

$$\begin{cases} A_{if} = \dfrac{4500V_{cf}}{C_d\sqrt{\frac{2(P_2-P_f)}{\rho}}} \\ A_{ir} = 4500V_{cr}C_d\sqrt{\frac{2(P_2-P_r)}{\rho}} \end{cases} \tag{4.14}$$

where: $Acf$ is the cross-sectional area of the front shaft inlet valve and $Acr$ is the cross-sectional area of the rear shaft inlet valve.

Similarly, the EHB discharge valve must have a specific pressure relief capacity under the decompression condition of emergency braking. Generally, the decompression rate of anti-lock braking is 100 MPa/s. The cross-sectional area of the corresponding discharge valve of the front and rear axle wheels is

$$\begin{cases} A_{of} = \dfrac{6000V_{cf}}{C_d\sqrt{\frac{2P_f}{3\rho}}} \\ A_{Or} = \dfrac{6000V_{cr}}{C_d\sqrt{\frac{2P_r}{5\rho}}} \end{cases} \tag{4.15}$$

where: $Aof$ is the cross-sectional area of the front shaft discharge valve; $Aor$ is the cross-sectional area of the rear shaft discharge valve.

## 4.2 Energy-Saving Design of Integrated Brake System for Electromagnetic and Friction

### *4.2.1 Energy Consumption Analysis of Electro-Hydraulic Brake*

#### 4.2.1.1 Modeling of Energy Consumption of Electro-Hydraulic Brake

Modeling of Accumulator Discharge Process

The following formula expresses the change of hydraulic energy when the accumulator stores or releases high-pressure brake fluid during braking.

$$W_A = -\int_{V_{A0}}^{V_{A1}} P_A dV_A \tag{4.16}$$

where: $V_{A1}$—the gas volume of accumulator after braking, m$^3$.

According to Boyle's law of Thermodynamics:

$$P_0 V_0^n = P_1 V_1^n = P_2 V_2^n \tag{4.17}$$

where: $P_0$—pre-charging pressure of accumulator, Pa; $V_0$—Effective volume of the accumulator, m$^3$; $P_1$—Maximum working pressure of the system, Pa; $P_2$—Minimum working pressure of the system, Pa; $V_1$—The gas volume of accumulator under maximum pressure, m$^3$; $V_2$—The gas volume of the accumulator at minimum pressure, m$^3$.

The expression of the adequate volume of the accumulator is as follows:

$$V_0 = \frac{\Delta V}{P_0^{0.715}\left[\left(\frac{1}{P_2}\right)^{0.715} - \left(\frac{1}{P_1}\right)^{0.715}\right]} \tag{4.18}$$

where: $\Delta V$—the adequate volume of the accumulator, $\Delta V = V_2 - V_1$, m$^3$.

Assuming that the discharge process of the accumulator is adiabatic, n is taken as 1.4, and Eq. (4.17) is substituted into Eq. (4.16) for simplification to obtain the energy released during the discharge process of the accumulator

$$W_A = P_0 V_0^n \frac{V_{A1}^n - V_{A0}^n}{n-1} \tag{4.19}$$

From Eqs. (4.18) and (4.19), it can be seen that in the process of automobile braking, the energy released by the accumulator is related to the volume change of the brake fluid of the accumulator, the design parameters of the accumulator (accumulator pre-charging pressure, accumulator effective displacement) and the system design parameters (system maximum working pressure and system minimum working pressure).

Modeling of Accumulator Charging Process

When the gas pressure of the accumulator is lower than the minimum working pressure of the system, the motor pump starts to work. It replenishes high-pressure brake fluid to the accumulator until the actual gas pressure in the accumulator is equal to the maximum working pressure of the system. In this process, the following formula can express the energy consumed by the motor pump:

$$W = P_{pump} \cdot t_P = \frac{P_P \times Q_P}{60\eta_P} \cdot t_P \tag{4.20}$$

where: $P_{pump}$—motor pump driving power, W; $t_P$—Charging time of accumulator, s; $\eta_P$—Motor pump efficiency.

The following formula can calculate the charging time of the accumulator:

$$t_P = \varepsilon \frac{V_2 - V_1}{Q_P} \tag{4.21}$$

where: $\varepsilon$—leakage coefficient of a hydraulic pump.

Substituting Eq. (4.21) into Eq. (4.20) can obtain the energy consumed by the accumulator during liquid charging:

$$W = \varepsilon \frac{P_P(V_2 - V_1)}{60\eta_P} \tag{4.22}$$

From the analysis of Eqs. (4.19) and (4.22), we can see that the electric energy consumed by the accumulator during liquid filling is converted into hydraulic energy for storage, and the stored hydraulic energy is released during fluid discharge. The greater the effective displacement of the accumulator, the higher the maximum working pressure of the system, and the more electric energy the accumulator consumes in filling liquid. Therefore, in the design process, we should reduce the effective displacement of the accumulator and the maximum working pressure of the system to reduce the energy consumed by the single charge of the accumulator.

Modeling of Brake Fluid Requirement

Be based on the analysis results in Sect. 4.2.1.1, the amount of energy released during the accumulator discharge is related to the change in the volume of brake fluid in the accumulator. The shift in accumulator brake fluid volume is directly related to the brake fluid demand. Thus, it is indispensable to establish and analyze the mathematical model of brake fluid demand.

Considering that the pressure in the brake wheel cylinder changes slowly in the braking idle stroke stage and ignoring the dynamic influence of hydraulic pressure, it is assumed that:

$$\begin{cases} \dfrac{d^2 X_C}{dt^2} = 0 \\ \dfrac{dX_C}{dt} = 0 \end{cases} \tag{4.23}$$

According to the above formula, multiply both sides of the procedure (3.14) by the cross-sectional area of the brake wheel cylinder piston to obtain the following formula.

$$P_C A_C^2 = (K_C + K_s) V_{C0} \tag{4.24}$$

The fluid demand of the brake wheel cylinder in the braking idle stroke stage is:

$$V_{CE} = \frac{P_C A_C^2}{K_C + K_s} \tag{4.25}$$

The simultaneous derivation of both sides of Eq. (4.25) shows that the liquid demand of brake wheel cylinder in the stable pressurization stage is:

$$V_{CW} = \int_{P_{C0}}^{P_{C1}} \frac{V_C}{K_C} \frac{dP_C}{dt} = \frac{V_C}{K_C}(P_{C1} - P_{C0}) \tag{4.26}$$

where: $P_{CO}$, $P_{CI}$—the pressure of brake wheel cylinder before and after braking, Pa.

Integrate the two ends of Eq. (3.12) to obtain that the liquid demand of the brake pipeline in the stable pressurization stage is:

$$V_{HW} = \frac{V_{H0}}{K_f}(P_{hi} - P_{ho}) \tag{4.27}$$

where: $V_{HO}$—the volume of brake pipeline under initial state, m$^3$.

Automobile brake pipeline includes rigid brake pipe and brake hose. Generally, we do not consider the influence of elastic deformation of frozen brake pipe; The radial expansion of the brake hose during braking is significant, and we cannot directly ignore it. Therefore, the bulk elastic modulus of brake fluid in formula (4.27) can no longer reflect the fluid flow characteristics inside the brake pipeline, and its value shall be modified to the equivalent bulk elastic modulus of brake fluid and brake pipeline elasticity:

$$K_{eq} = \frac{K_H E_s b_s}{E_s b_s + K_H d_s} \tag{4.28}$$

where: $b_s, d_s$—inner diameter and wall thickness of brake hose, m; $E_s$—Longitudinal elastic coefficient of brake hose; $K_H$—Equivalent spring stiffness of brake pipeline, N/m.

According to the comprehensive formulas (4.25), (4.26), and (4.27), the volume of brake fluid required during braking is:

$$V_r = V_{CE} + V_{CW} + V_{HW} \tag{4.29}$$

According to the Eq. (4.29) analysis, the volume of brake fluid required in the braking process is closely relevant to the brake wheel cylinder pressure (break strength required by the driver) and the brake wheel cylinder in the piston area. That is, their values directly affect the energy consumption characteristics of electro-hydraulic braking.

#### 4.2.1.2 Analysis of Energy Consumption of Electro-Hydraulic Brake

It is tough to directly test and analyze the influencing factors of energy consumption characteristics of electro-hydraulic brakeing, which has complex test operations and high costs. Therefore, using the established rapid control prototype platform of integrated brake system for electromagnetic and friction can verify the effectiveness of the mathematical model of electro-hydraulic braking energy consumption proposed in Sect. 4.2.1. We can use the pressure sensor to collect the brake fluid pressure at the accumulator outlet and then calculate the hydraulic energy change during accumulator discharge according to Eq. (4.19). As shown in Fig. 4.1 and Fig. 4.2, after five boost and depressurization tests, the gas pressure in the accumulator decreases to about 14.6 MPa, and the energy in the accumulator decreases from 700 to 400 J; At the same time, the simulation curves of accumulator air pressure and braking energy consumption are in good agreement with the test curves. The above shows the effectiveness of the mathematical model of electro-hydraulic braking energy consumption. The mathematical model can be used to analyze the influencing factors of electro-hydraulic braking energy consumption characteristics.

To sum up, six main factors affect the energy consumption characteristics of electro-hydraulic brake: driver's required braking strength, the diameter of brake wheel cylinder piston, accumulator charging pressure, accumulator effective displacement, system maximum working pressure, and system minimum operating pressure. The following analyzes the impact of these parameters on the energy consumption characteristics of electro-hydraulic brake. The simulation parameters

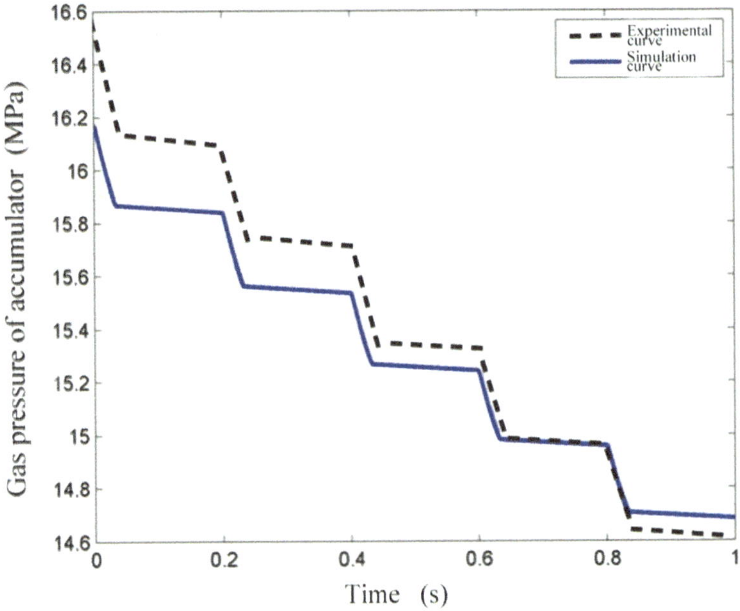


**Fig. 4.1** Simulation curve and test curve of the accumulator gas pressure

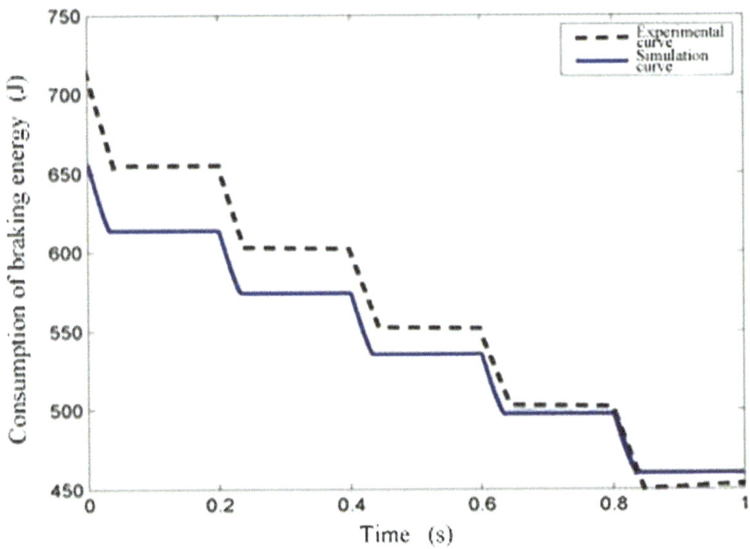


**Fig. 4.2** Simulation curve and test curve of the braking energy consumption

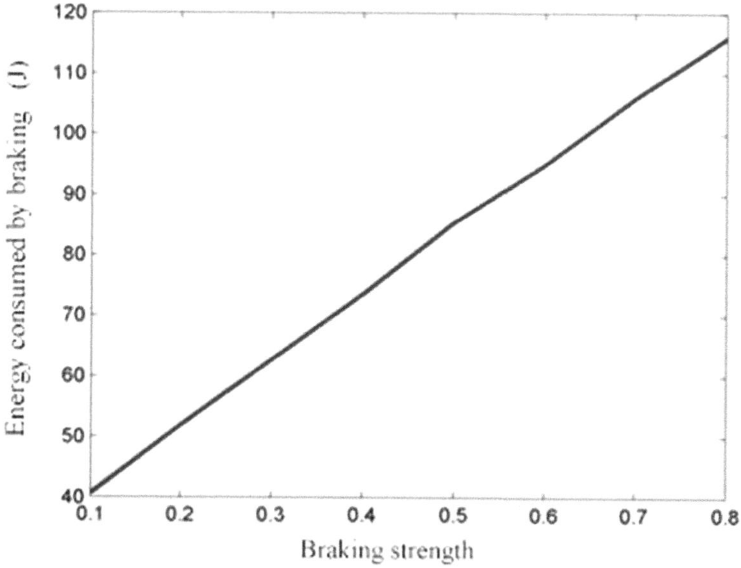


**Fig. 4.3** Curve of the EHB's energy dissipation changing with braking intensity

are set as follows: the required braking strength of the driver is 0.8, the maximum working pressure of the system is 16 MPa, the minimum working pressure of the system is 12 MPa, the charging pressure of the accumulator is 7.5 MPa, the displacement of the accumulator is 43 ml, and the piston diameters of the front and rear brake wheel cylinders are 60 mm and 42 mm respectively.

As shown in Fig. 4.3, when other simulation parameters remain unchanged, the braking energy consumption of electro-hydraulic braking increases linearly with the driver's braking intensity. When the braking strength increases from 0.1 to 0.8, the braking energy consumption increases from 40 J to about 117 J, and the braking strength has a significant impact on the braking energy consumption.

As Fig. 4.4 shows, the diameter of the brake wheel cylinder piston is in direct proportion to the braking energy consumption of electro-hydraulic brake. Therefore, reducing the diameter of the brake wheel cylinder piston can effectively mitigate the braking energy consumption. Still, the reduction of the diameter of the brake cylinder piston will cause the minimum working pressure of the system to be too high, which may further push up the maximum working pressure of the system. It makes the energy consumption characteristics of electro-hydraulic brake worse. In addition, higher system minimum pressure will also increase the cost of electro-hydraulic brake, and there is a risk of brake fluid leakage, which will affect safety.

From Figs. 4.5 and 4.6, it can be seen that the energy consumption of electro-hydraulic brake is not sensitive to changes in the effective displacement of the accumulator and the charging pressure. The selection of effective displacement of the

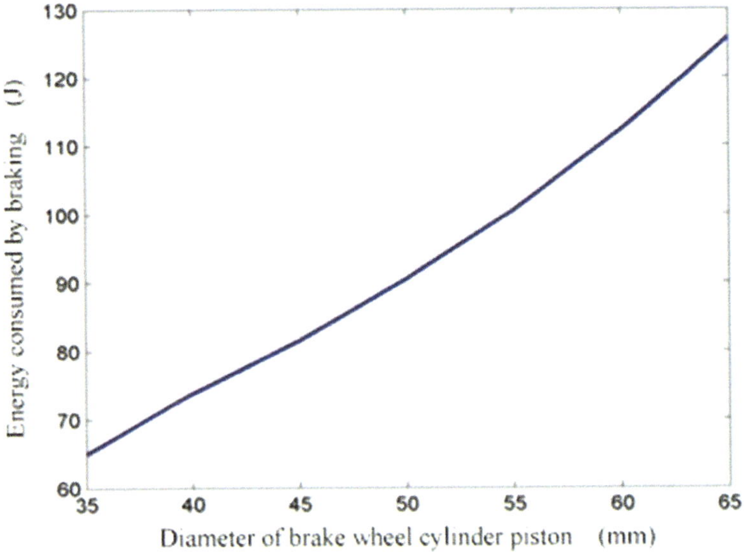


**Fig. 4.4** Curve of the EHB's energy dissipation changing with the diameter of brake wheel cylinder piston

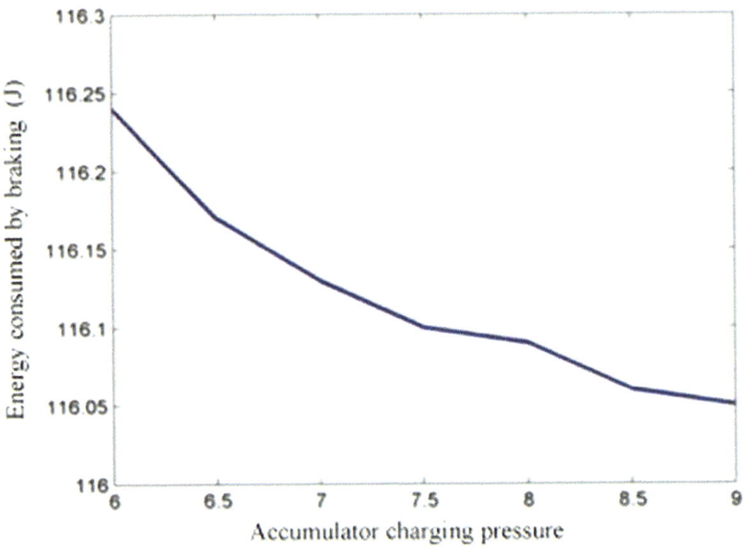


**Fig. 4.5** Curve of the EHB's energy dissipation changing with accumulator charging pressure

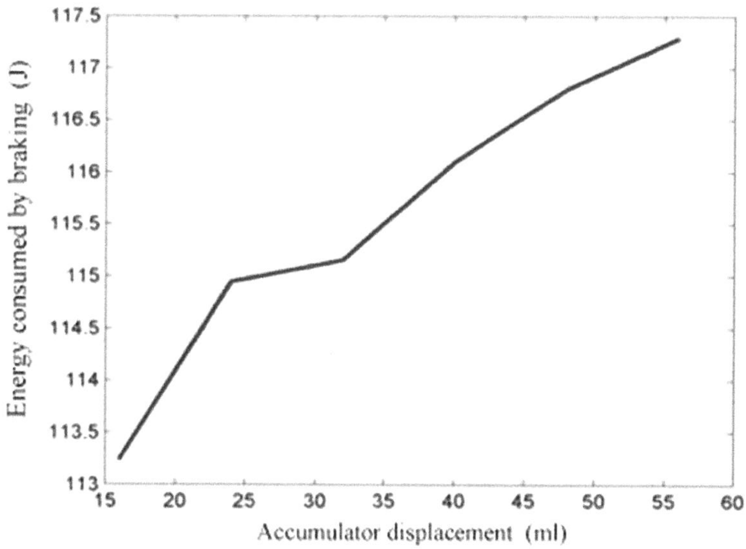


**Fig. 4.6** Curve of the EHB's energy dissipation changing with accumulator effective displacement

accumulator is affected by many factors. These factors include the need for the accumulator to provide several high-strength braking in case of motor failure and the need to limit the increase of the adequate volume of the accumulator. We should comprehensively consider these factors in the design process, and the influence of the effective displacement of the accumulator on the total energy consumption of electro-hydraulic brake can be ignored. The design of accumulator charging pressure is relatively random. The method of accumulator charging pressure is somewhat arbitrary. We mainly hope to reduce the practical volume of the accumulator by reasonably matching the system working pressure parameters.

As Figs. 4.7 and 4.8 show, the setting of system working pressure has a limited influence on the energy consumption of the accumulator during braking. When the maximum working pressure of the system increases from 15 MPa to 18 MPa, the energy consumption of electro-hydraulic brake increases by 10 J; The minimum working pressure of the system increases from 10 MPa to 13Mpa, and the energy consumption of electro-hydraulic brake increases by only about 4 J. Considering that the rise in the maximum working pressure of the system will significantly increase the driving power of the motor pump, it is not advisable to exceed the maximum working pressure of the system.

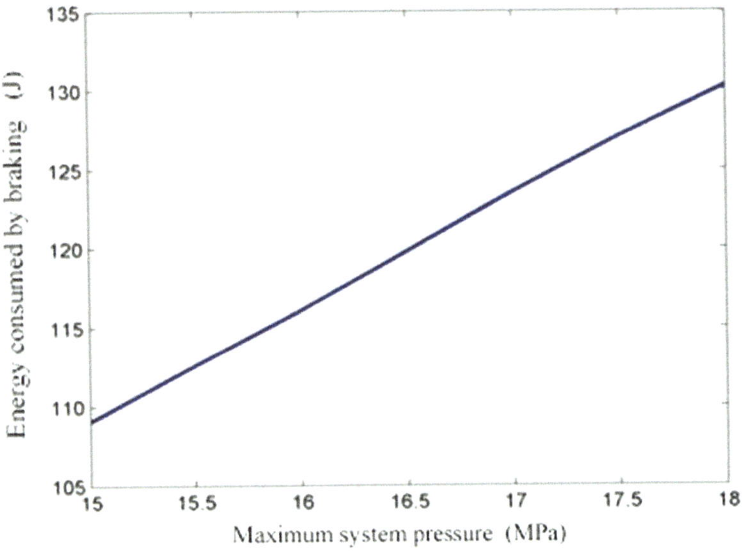


**Fig. 4.7** Curve of the EHB's energy dissipation changing with maximum system pressure

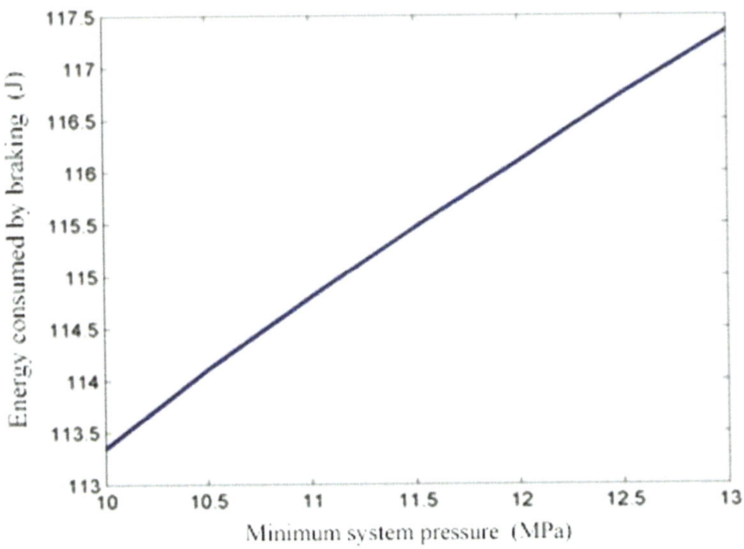


**Fig. 4.8** Curve of the EHB's energy dissipation changing with minimum system pressure

### 4.2.2 *Energy Consumption Analysis of Electromagnetic Brake*

#### 4.2.2.1 Modeling of Energy Consumption of Electromagnetic Brake

As Fig. 4.9 shows, the excitation winding of the electromagnetic brake principle prototype is fixed on the stator support, and each two series excitation windings are specified in the radial direction on the left and right sides of the brake disc. At the same time, multiple parallel excitation windings are fixed along the circumferential direction on the same side of the brake disc. Therefore, the calculation expression of electromagnetic braking energy dissipation power is:

$$P = 4N_p I^2 R_0 \tag{4.30}$$

where: $R_0$—resistance of single excitation coil, Ω.

1. Brake caliper 2. Brake disc 3. Excitation winding 4. Stator support

**Fig. 4.9** Prototype of the frictional and electromagnetic integrated brake of passenger car

According to Eq. (4.30), when the structural parameters of the electromagnetic brake are determined, the electromagnetic braking torque is related to the magnitude of the excitation current in the excitation coil and the number of energized excitation coils. The electromagnetic brake controller controls the excitation current by changing the conduction time between the excitation coil and the battery through the DC chopper. Simplify Eq. (3.22) to obtain the electromagnetic braking torque required to be output by the controller in the braking process as follows:

$$T_e = \frac{N_p S_p^2 \Delta_h \omega_d^2}{8\pi \rho_d k_a} B^2 \tag{4.31}$$

Of which:

$$B = \frac{8\pi \rho_d \mu_0 N I}{16\pi \rho_d l_g + \sqrt{2} K_e S_p \mu_0 \Delta_h \omega_d}$$

where: $K_e$—eddy current conversion coefficient.

The required excitation current is derived from the combined formula (4.30) and formula (4.31) as follows:

$$I = \sqrt{\frac{8\pi \rho_d k_a T_e}{N_p S_P^2 \Delta_h \omega_d^2}} \left( \frac{16\pi \rho_d l_g + \sqrt{2} K_e S_p \mu_0 \Delta_h \omega_d}{8\pi \rho_d \mu_0 N} \right) \tag{4.32}$$

The resistance of a single excitation coil can be expressed as follows:

$$R_0 = \rho_w \frac{l_w}{s} \tag{4.33}$$

Of which:

$$l_w = n_2 \times [n_1 (d + 2\delta)\pi + n_1 (n_1 - 1)\pi d_0]$$

$$s = \frac{1}{4} \pi d_0^2$$

$$n_1 = fix\left( \frac{D - d - 2\delta}{2d_0} \right)$$

$$n_2 = fix\left( \frac{h - 2\delta}{d_0} \right)$$

where: $l_w$—length of copper conductor, m; s—Cross-sectional area of copper conductor, $m^2$; $n_1$—Number of coil radial turns; $n_2$—Coil axial turns; D—Outer diameter of coil frame, m; δ—Coil frame thickness, m; $d_0$—Copper wire diameter, m; h—Width of coil frame, m.

According to the above formula (4.33), the resistance of a single excitation coil is related to the excitation coil parameters (coil skeleton outer diameter, coil skeleton length, coil skeleton thickness, and iron core diameter) and copper conductor parameters (copper conductor diameter and copper conductor resistivity).

According to formula (4.33), the calculation formula of the number of turns of the excitation coil can be obtained as follows:

$$N = n_1 \cdot n_2 \tag{4.34}$$

Combined formulas (4.30)–(4.34), power consumed by electromagnetic brake is relevant to air gap length, excitation coil parameters (coil skeleton width, coil skeleton outer diameter, coil skeleton thickness, iron core diameter), copper conductor parameters (copper conductor diameter, copper conductor resistivity), brake disc parameters (brake disc resistivity, brake disc relative permeability), brake disc angular velocity and required braking torque.

#### 4.2.2.2 Analysis of Energy Consumption of Electromagnetic Brake

It is also complicated to directly test and analyze the influencing factors of electromagnetic braking energy consumption characteristics. So we use the built rapid control prototype platform of integrated brake system for electromagnetic and friction to verify the effectiveness of the electromagnetic braking energy consumption mathematical model proposed in Sect. 4.2.1.1. As shown in Fig. 4.10, the test value of electromagnetic braking energy consumption characteristics can be in good agreement with the simulation curve. The mathematical model of electromagnetic braking energy consumption proposed in this paper is accurate. This mathematical model can be used to investigate the influencing factors of electromagnetic braking energy consumption characteristics.

According to the combined Eqs. (3.19) and (3.20), the temperature of the excitation coil has a specific impact on the resistivity of the copper conductor, and the temperature of the brake disc will also affect the values of the resistivity and relative permeability of the brake disc; The thickness of the coil frame is directly relevant to the width and outer diameter of the coil frame. So we can ignore the influence of coil skeleton thickness; Urban vehicle driving conditions will directly impact the brake disc angular velocity and required braking torque of the electromagnetic brake. To sum up, we will analyze the energy consumption characteristics of electromagnetic brake from the following five perspectives: air gap length, copper wire diameter, coil frame width, outer diameter, iron core diameter, excitation coil, and brake disc temperature urban vehicle driving conditions. The setting of stimulation parameters is the same as that of the electromagnetic brake principle prototype: air gap length is 1 mm, copper wire diameter is 1.8 mm, coil frame width is 58 mm, coil frame outer diameter is 86 mm, the iron core diameter is 54 mm, coil frame thickness is 2 mm, and brake disc center radius is 120 mm.

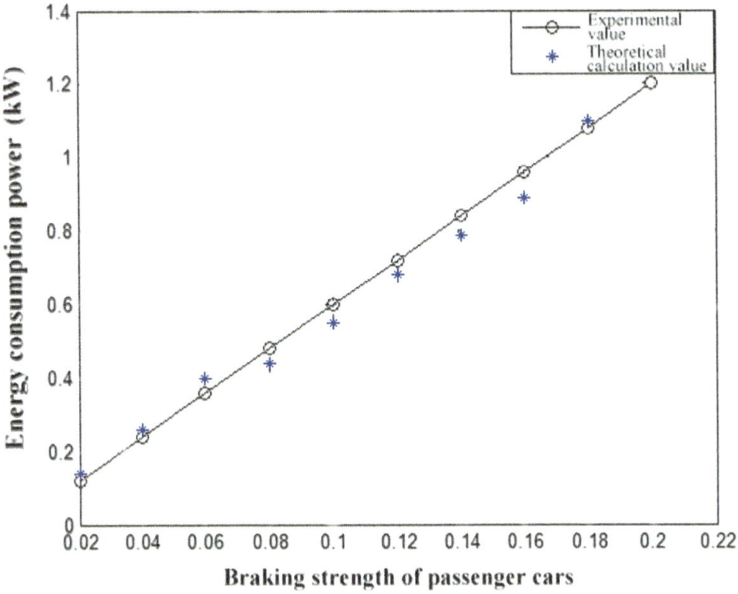


**Fig. 4.10** Simulation curve and test curve of the ECB's energy-consumption power

Next, we will investigate the influence of air gap length on the energy consumption characteristics of electromagnetic brake. Based on Eq. (4.32), when the target braking torque of electromagnetic brake is the same, the larger the air gap length of the electromagnetic brake, the greater the excitation current required. As shown in Fig. 4.11, the air gap length will be linearly proportional to the electromagnetic energy consumption power. When the air gap length increases from 0.7 mm to 1.4 mm, the electromagnetic energy consumption power will more than double, that is, from 0.4 kW to 0.9 kW. Therefore, we can think that the air gap length will significantly impact the energy dissipation power of electromagnetic brake. Thus, in the process of design, production, and assembly of electromagnetic brake, we should minimize the design value of air gap length and accurately ensure that the air gap length meets the design requirements, which is beneficial to reducing the energy consumption power brake.

As shown in Fig. 4.12, when the copper wire diameter changes from 1.6 mm to 2.3 mm, the changing trend of electromagnetic energy dissipation power fluctuates with the increase of copper wire diameter. However, the value of electromagnetic energy dissipation power does not increase significantly. That is, the influence curve of copper wire diameter on electromagnetic energy dissipation power fluctuates slightly. This phenomenon is because the copper wire diameter will directly impact the resistance and turns of the excitation coil. Under a specific coil frame size, if the copper wire diameter continues to increase, the resistance of the excitation coil will decrease. The electromagnetic energy dissipation power will reduce

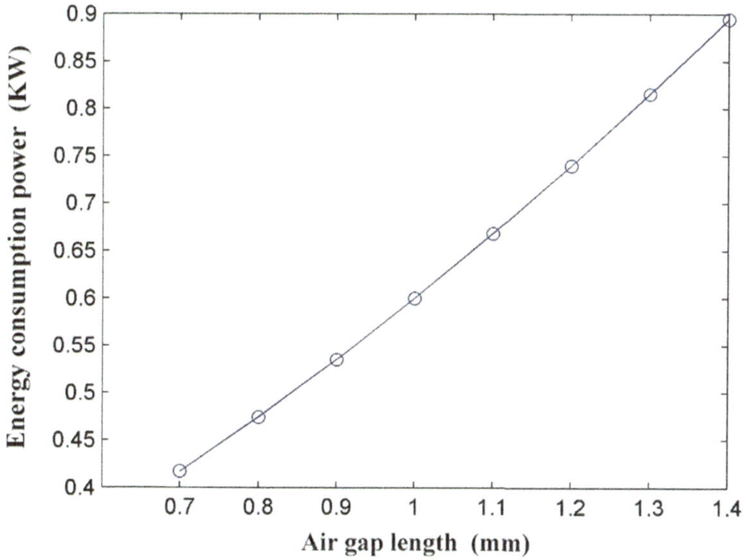


**Fig. 4.11** Impact of the gas length on ECB's energy-consumption power

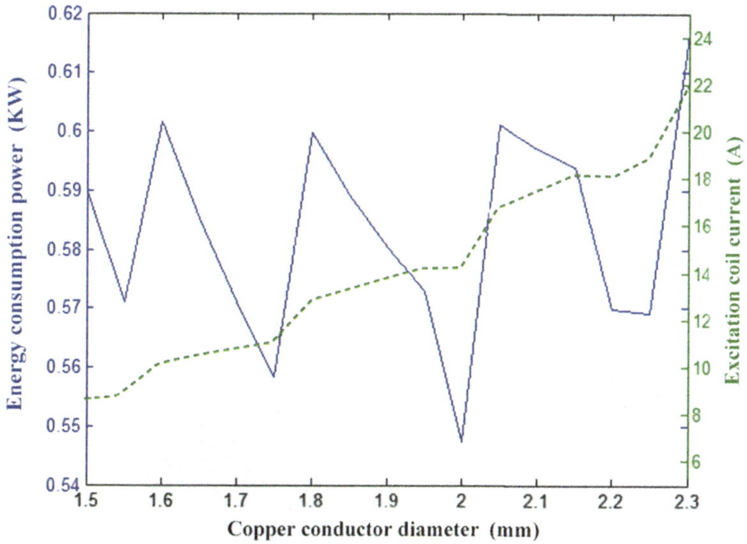


**Fig. 4.12** Impact of the diameter of copper conductor on ECB's energy-consumption power

according to Eqs. (4.30) and (4.33). According to Eqs. (4.32) and (4.34), when the number of turns of the excitation coil decreases and the current of the excitation coil increases, the electromagnetic energy dissipation power will increase. However, due to the small order of magnitude of copper wire diameter and the mutual cancellation of positive and negative effects on electromagnetic energy consumption power, the influence of copper wire diameter on electromagnetic energy consumption power is fragile. As shown in Fig. 4.12, with the increase of copper wire diameter, the excitation current increases rapidly from 9A to 22a, indicating that the maximum current density is sensitive to the change of copper wire diameter. Therefore, in the initial stage of electromagnetic brake design, we can ignore the influence of copper wire diameter on the energy consumption characteristics of an electromagnetic brake. After the electromagnetic brake is designed, we can check the copper wire diameter with the constraint of maximum current density. After the electromanetic brake design is completed, we can check the copper wire diameter with maximum current thickness.

As shown in Fig. 4.13, with the increase of the width and outer diameter of the coil skeleton, the electromagnetic energy dissipation power will continue to decrease. The influence curve of the width and outer diameter of the coil skeleton on the electromagnetic energy dissipation power is piecewise stepped, which is caused by the rounding function (x) when solving the radial and axial turns of the excitation coil in Eq. (4.33). Combining formula (4.33) and formula (4.34), we can find that the increase of coil frame width and outer diameter will increase the number of turns of an excitation coil and the resistance of the excitation coil. Still, the influence relationship between the number of turns of the excitation coil and the resistance of the excitation

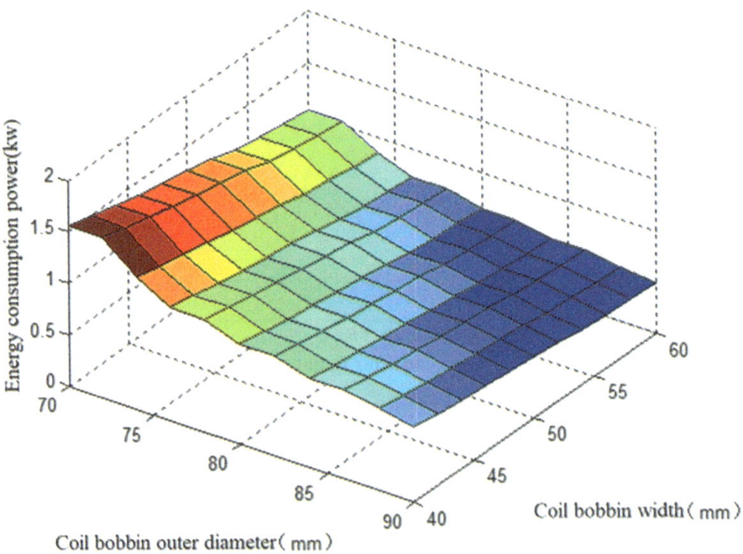


**Fig. 4.13** Impact of the width and diameter of coil former on ECB's energy-consumption power

coil on electromagnetic energy dissipation power is opposite. According to Eq. (4.30), the increase of excitation coil resistance will increase the electromagnetic energy dissipation power; According to Eq. (4.32), the electromagnetic energy dissipation power will decrease with the rise in the number of turns of the excitation coil. The influence of the two on energy consumption power becomes very fuzzy. The main reason for the phenomenon, as shown in Fig. 4.13, is that the number of turns of the excitation coil and the resistance of the excitation coil has different proportions on the electromagnetic energy dissipation power, and the number of turns of the excitation coil has a relatively large proportion.

As shown in Fig. 4.14, the electromagnetic energy dissipation power does not increase significantly when the core diameter increases from 50 to 57 mm. The electromagnetic energy dissipation power rises significantly when the core diameter increases from 57 to 65 mm—combined Eqs. (4.30)–(4.34) explain the above phenomena. The rise in iron core diameter will reduce the number of turns and resistance of the excitation coil, and the relationship between the number of turns and resistance of excitation coil on electromagnetic energy dissipation power is fuzzy. At the same time, we can find that the increase of core diameter will also lead to the rise of the eddy current area. The area of the eddy current action area is inversely proportional to the excitation current. The larger the eddy current action area is, the smaller the electromagnetic energy dissipation power is. Therefore, we can think that when the core diameter is between 50 and 57 mm, the increase of core diameter will lead to the rise in eddy current action area, and its positive effect on the reduction of electromagnetic energy consumption power will be dominant. When the core diameter is greater than 57 mm, the number of turns and the resistance of the excitation

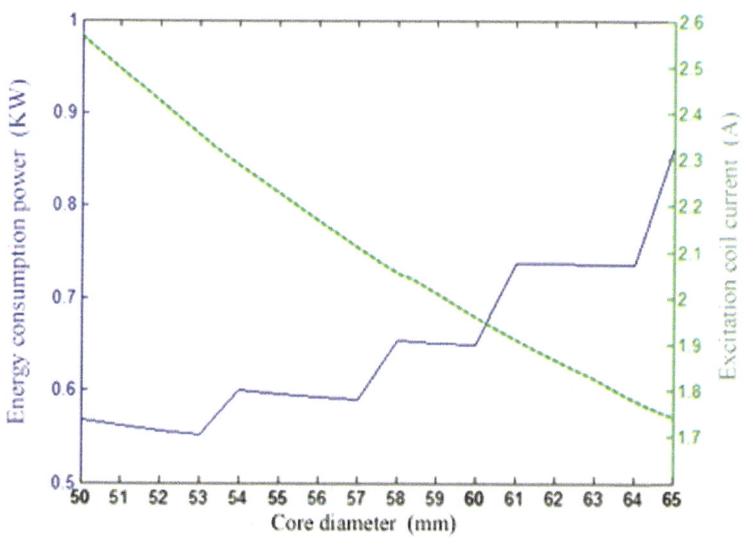


**Fig. 4.14** Impact of the diameter of iron core on ECB's energy-consumption power

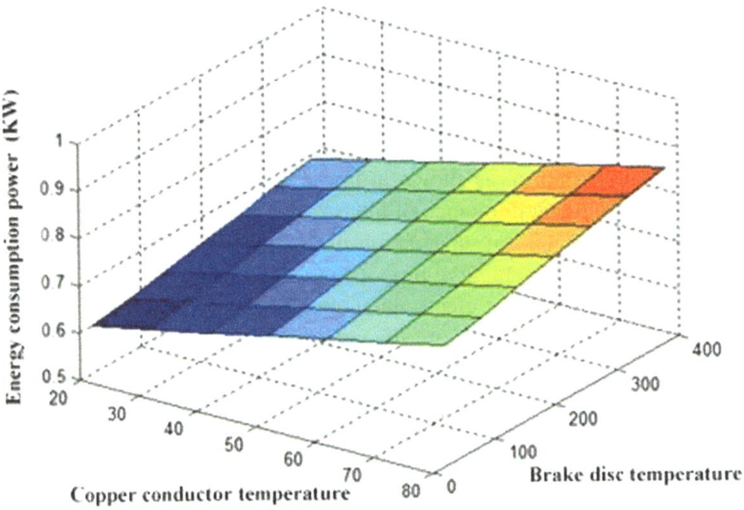


**Fig. 4.15** Impact of the brake disc temperature on ECB's energy-consumption power

coil decreases with the increase of the core diameter. Its negative effect on the rise in electromagnetic energy dissipation power will dominate.

As shown in Fig. 4.15, under the same initial state, when the copper conductor temperature is increased from 20 °C to 80 °C or the brake disc temperature is increased from 20 °C to 400 °C, the electromagnetic braking energy consumption power will increase by about 0.1 kW. We can find that the copper conductor temperature and brake disc temperature directly proportioned to the electromagnetic energy consumption power. Considering that the brake disc has good heat dissipation performance and temperature stability, in the design process of electromagnetic brake, the influence of brake disc temperature on electromagnetic braking energy dissipation power is ignored. The ventilation and heat dissipation of the excitation coil is improved by optimizing the results of stator support to reduce electromagnetic braking energy dissipation power.

As shown in Fig. 4.16, the electromagnetic energy consumption power is immense when the vehicle speed is low. With the vehicle speed increase, the capacity first decreases and then increases slowly with a slight slope. When the vehicle speed is 50 km/h, the electromagnetic energy consumption power reaches the minimum. When the vehicle speed is within the range of 30 km/h to 80 km/h, the electromagnetic braking energy consumption power of the electromagnetic brake principle prototype is relatively low. Therefore, we call this vehicle speed range "low energy consumption power area." When the vehicle speed is 10 km/h, the electromagnetic energy consumption power is greater than the minimum value of electromagnetic energy consumption power. The electromagnetic energy consumption power will increase significantly when the vehicle speed is less than 10 km/h. Therefore, electromagnetic brake should be prohibited when the vehicle speed is less than 10 km/h. Comparing

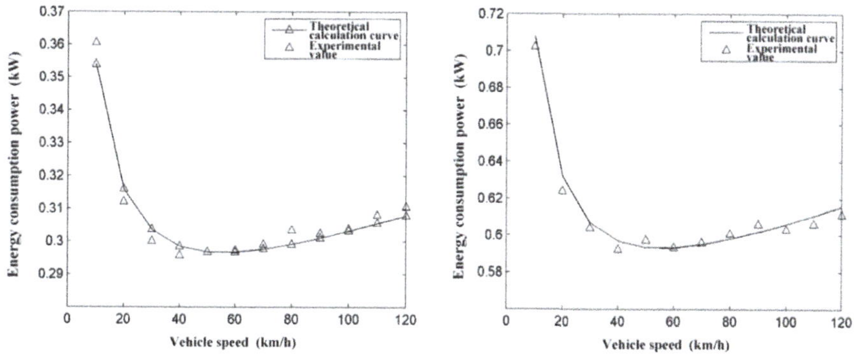


(a) Z=0.05 (b) z=0.1

**Fig. 4.16** Impact of vehicle running speed on energy-consumption power

Fig. 4.16a, b, we can discover that the changing tendency of electromagnetic braking energy depletion power with vehicle speed is independent of vehicle demand braking strength.

Figure 4.17a shows that the ECE + EUDC and UDDS cycle is typical urban vehicle driving conditions. Under the ECE + EUDC process, the average breaking strength is 0.079, and the maximum breaking strength is 0.139. As shown in Fig. 4.17b, under UDDS cycle conditions, the average breaking strength is 0.058, and the whole breaking strength is 0.148; The initial braking speed range of both is 20 km/h-40 km/h, and the average driving speed is about 32 km/h. Compared with Figs. 4.16 and 4.17, the "low energy consumption power area" of the electromagnetic brake principle prototype overlaps less with the braking initial speed range of two representative urban driving situations, so the energy consumption characteristics of the electromagnetic brake principle prototype have significant room for improvement. Based on the accurate dynamic mathematical model established in Sect. 3.2,

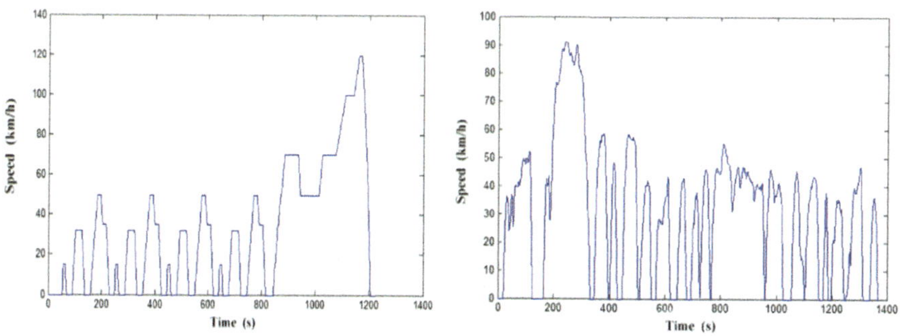


(a) ECE+EUDC cycle condition (b) UDDS cycle condition

**Fig. 4.17** Urban driving cycle of vehicles

we can optimize the design of the electromagnetic brake from the perspective of making the "low energy consumption power area" of the electromagnetic brake cover the range of initial braking speed under representative urban driving situations as much as possible, which is conducive to reducing the power consumption of the electromagnetic brake in actual use.

### *4.2.3 Energy-Saving Design of Integrated Brake System for Electromagnetic and Friction*

#### 4.2.3.1 Design Variable

According to the above analysis, the brake parameters (the diameter of the brake cylinder piston), accumulator design parameters (inflation pressure, effective displacement), and system design parameters (system maximum pressure and system minimum pressure) have an impact on the energy consumption characteristics of electro-hydraulic brake; In electromagnetic brake, the parameters of excitation coil (outer diameter of coil frame, length of coil frame, core diameter, wire diameter of copper conductor) and brake parameters (central radius of brake disc) are directly related to the energy dissipation power of electromagnetic brake. In the meanwhile, it should be noted that an integrated brake system for electromagnetic and friction has the merit of independently controlling the braking torque of four wheels. The braking force of the front and rear brakes of the vehicle does not need to set a fixed ratio. It can be distributed based on the ideal braking force distribution curve to ensure the vehicle's safety. Therefore, in the design process, it is no longer necessary to achieve this goal by optimizing and matching the friction brake like the traditional braking system. Therefore, these ten parameters can be used as design variables to establish the energy-saving optimization mathematical model of an integrated brake system for electromagnetic and friction.

#### 4.2.3.2 Objective Function

Considering the complexity of energy consumption characteristics of electro-hydraulic brake and electromagnetic brake, we should set independent objective functions for electro-hydraulic brake and electromagnetic brake. Therefore, the energy-saving optimization design of integrated brake system for electromagnetic and friction is a multi-objective and multivariable optimization design. The calculation expression of energy consumption of electro-hydraulic brake under the ECE + EUDC cycle is as follows:

$$E_{EHB} = floor(d_1) \times W + [b_1 - floor(d_1)] \times b_2 \tag{4.35}$$

Of which:

$$\begin{cases} d_1 = \dfrac{\sum_{i=1}^{j} V_{A1}^n - V_{A0}^n}{V_2^n - V_1^n} \\ d_2 = P_0 V_0^n \dfrac{V_2^n - V_1^n}{n-1} \end{cases}$$

where: *floor(x)*—rounding function;

In electro-hydraulic brake, the airbag accumulator is integrated with the hydraulic regulator and installed in the engine compartment, which requires the accumulator to be as small as possible. We ignore the volume of the accumulator shell and assume that the gas volume in the accumulator is the accumulator volume. That is, the maximum accumulator volume is the adequate volume of the accumulator in the initial state.

Considering the difference of importance between the two indexes and assigning weighting factors to the two evaluation indexes, the objective function of energy-saving optimization design of electro-hydraulic brake is as follows:

$$f_H(x) = \gamma_1 E_{EHB} + \gamma_2 V_0 \tag{4.36}$$

where: $\gamma_1$—weighting factor of energy consumption of electro-hydraulic brake; $\gamma_2$—Accumulator volume weighting factor.

Similarly, we investigate the electric energy consumed by electromagnetic brake under the ECE-EUDC cycle as an energy-saving index.

$$E_{ECB} = 4N_p R_0 \int_{\omega_1}^{\omega_2} I^2(\omega) d\omega \tag{4.37}$$

where: $\omega_1$—wheel angular velocity at the beginning of single braking, rad/s; $\omega_2$—Wheel angular velocity at the end of single braking, rad/s.

The increase of unsprung mass caused by the installation of electromagnetic brake affects the vehicle's dynamic performance and increases the energy consumption in the car's operation. To measure the advantages and disadvantages of electromagnetic brake in lightweight, we require that the total mass of electromagnetic brake should be reduced. The entire group of the electromagnetic brake can be expressed as follows.

$$m = m_d + m_h + 2N_p(m_i + m_w) \tag{4.38}$$

where: $m_d$—brake disc mass, kg; $m_h$—Mass of stator support, kg; $m_i$—Core mass, kg; $m_w$—Mass of copper conductor, kg.

By assigning weighting factors to the above two evaluation indexes of electromagnetic brake, the objective function of energy-saving optimization design of electromagnetic brake is obtained as follows:

$$f_E(x) = u_1 E_{ECB} + u_2 m \tag{4.39}$$

where: $u_l$—energy consumption weighting factor of electromagnetic brake; $u_2$—Weighting factor of electromagnetic braking quality.

#### 4.2.3.3 Constraint Condition

Constraints of Electro-Hydraulic Brake

(1) Maximum/minimum working pressure of the system: In case of emergency braking, when the electro-hydraulic brake is pressurized, the pressure change rate of the brake wheel cylinder shall meet the requirement that the pressure of the brake wheel cylinder can increase to the level of locking the wheels on the high attachment road within the expected pressurization time. Generally, the pressurization rate of vehicle anti-lock braking is required to be 75 MPa/s. This requires the maximum working pressure of the system. Too low the maximum working pressure of the system will lead to insufficient pressurization capacity of the anti-lock braking system. When the cross-sectional area of the inlet valve corresponding to the front and rear brake wheel cylinders is determined, the constraint conditions that the maximum working pressure of the system should meet according to the combined formula (4.28) are:

$$\begin{cases} C_d A_{cf}\sqrt{\dfrac{2(P_1 - P_{cf})}{\rho_f}} > \dfrac{4500V_{cf}}{P_{cf}} \\ C_d A_{cr}\sqrt{\dfrac{2(P_1 - P_{cr})}{\rho_f}} > \dfrac{4500V_{cr}}{P_{cr}} \end{cases} \tag{4.40}$$

The minimum working pressure of the system shall meet the following constraints:

$$P_2 > \max(P_{cf}, P_{cr}) \tag{4.41}$$

where: *max(x)*—maximum value function;

(2) Piston diameter of brake wheel cylinder: according to the above analysis, the piston diameter of the brake wheel cylinder shall be reduced as much as possible to reduce the system's total energy consumption. However, to prevent adverse effects such as brake fluid leakage caused by excessive pressure in the brake system, the traditional hydraulic braking system requires that the pressure of the front brake wheel cylinder shall not exceed 7–8 MPa, and the pressure of the rear brake wheel cylinder shall not exceed 11–12 MPa. Therefore, when the vehicle model parameters are determined, the force of front and rear brake wheel cylinders shall not exceed 11 MPa. The constraint conditions for the piston diameter of the front and rear brake wheel cylinders are:

$$\begin{cases} d_{cf} > \sqrt{\dfrac{G\varphi(b+\varphi h_g)R_{roll}}{11\pi \cdot \eta_C(a+b)r_m}} \\ d_{cr} > \sqrt{\dfrac{2G\varphi(a-\varphi h_g)R_{roll}}{11\pi \cdot \eta_C(a+b)r_m}} \end{cases} \tag{4.42}$$

(3) Accumulator inflation pressure: There are two methods to prevent the accumulator from leaking or shortening its service life due to excessive deformation of the accumulator's bladder during electro-hydraulic braking. On the one hand, the accumulator bladder cannot be over-compressed when the system pressure increases; On the other hand, when the system pressure decreases, the accumulator skin bag cannot be overextended, and the constraints that the accumulator inflation pressure should meet are:

$$0.25P_1 < P < 0.9P_2 \tag{4.43}$$

(4) Effective displacement of accumulator: When the motor pump fails, the accumulator shall be able to complete high-strength braking at least 4–5 times, which requires that the effective displacement of the accumulator should be greater than the volume of brake fluid required for five times of vehicle high-intensity braking. The constraints that the effective removal of the accumulator shall meet are:

$$\Delta V > 5 \times \sum_{i=1}^{4} V_i \tag{4.44}$$

where: $V_i$—liquid consumption of each braking circuit during braking, $m^3$.

Constraints of Electromagnetic Brake

(1) Maximum current: The excitation coil in the electromagnetic brake needs to be connected with a large current, which is limited by the automotive electrical system, so the constraint conditions for the wind that can be input by the excitation coil are as follows:

$$I \leq I_{\max} \tag{4.45}$$

where: $I_{max}$—a maximum discharge current of automobile generator, A.

(2) Air gap magnetic induction strength: The electromagnetic brake's iron core and brake disc are soft magnetic materials. To prevent redundancy in the magnetic circuit design of the electromagnetic brake, it is required that the design value of air gap magnetic induction strength should be less than or equal to the saturated magnetic induction strength of the soft magnetic material. The constraints met

by the air gap magnetic induction strength of the electromagnetic brake:

$$B \leq B_{\mathrm{m}} \tag{4.46}$$

where: $B_m$—saturated magnetic induction strength of soft magnetic material, T.

(3) Maximum current density: During the continuous operation of the electromagnetic brake, the copper wire in the excitation coil will be electrified for a long time, and too much heat will accumulate in the excitation coil and cannot be dissipated. Once the temperature of the excitation coil exceeds the maximum allowable temperature, the insulation of the excitation coil will be reduced, and even the tightly wound copper wires will be broken down. Therefore, the current density of copper conductor is required to meet the following constraints:

$$J \leq \frac{4I}{\pi d_0^2} \tag{4.47}$$

(4) Outer diameter and width of coil frame: The width of the coil frame is limited by the space of vehicle wheels. The constraints of the outer diameter of coil frame and center radius of brake disc are as follows:

$$r_1 < r_m + \frac{d}{2} < r_2 \tag{4.48}$$

where: $r_l$—inner diameter of the brake disc, m; $r_2$—Outer diameter of the brake disc, m.

#### 4.2.3.4 Energy-Saving Design Example

This paper uses a genetic algorithm to solve the energy-saving optimization design of integrated brake system for electromagnetic and friction to obtain the optimal global solution. The penalty function method is used to deal with the constraints. Table 4.1 shows the comparison between the energy-saving optimization value of electro-hydraulic brake and the original parameter value.

As shown in Table 4.1, the maximum working pressure of the system decreases the most, which shows that reducing the maximum stress of the system is able to efficaciously decrease the electro-hydraulic brake's energy consumption. The minimum working pressure of the system decreases slightly, while the charging pressure of the accumulator increases significantly. Combined with the above analysis, it can be seen that the reason for the significant increase of accumulator inflation pressure is that it is conducive to reducing the adequate volume of the accumulator. The effective displacement of the accumulator is reduced to a certain extent, which is due to the reduction of the piston diameter of the front and rear wheel cylinders, which reduces the total liquid demand during braking and makes a specific contribution to reducing the volume of the accumulator. The energy-saving optimization results can

**Table 4.1** Optimal energy-saving values and original parameter values of EHB

| | Original parameters | Energy-saving optimization results |
|---|---|---|
| Maximum working pressure of the system/MPa | 16 | 14.4 |
| Minimum working pressure of system/MPa | 12 | 11.2 |
| Accumulator charging pressure/MPa | 7.5 | 9.8 |
| Effective displacement of accumulator/ml | 43 | 33 |
| Diameter of front brake wheel cylinder piston/mm | 60 | 56 |
| Diameter of Rear brake wheel cylinder piston/mm | 42 | 44 |
| Effective volume of accumulator/L | 0.339 | 0.221 |
| Energy consumption under 10 UDDS cycles/J | 7635.5 | 6222.3 |

significantly reduce the adequate volume of the accumulator. The practical importance of the accumulator with the original parameters is not within the reasonable design range, and the optimized accumulator volume is more affordable. The energy consumption of electro-hydraulic brake based on 10 UDDS cycles is also investigated in the table. The results show that the energy consumption based on 10 UDDS cycles is reduced compared with the original parameters.

The comparison between the optimized value of electromagnetic brake energy saving and the original parameter value is shown in Table 4.2. After optimization,

**Table 4.2** Energy-saving optimal values and the original values of ECB

| | Test principle and prototype parameters | Energy-saving optimization results |
|---|---|---|
| Coil frame width/mm | 58 | 54 |
| Outer diameter of coil frame/mm | 86 | 92 |
| Core diameter/mm | 54 | 58 |
| Center radius of the brake disc/mm | 120 | 123 |
| Copper wire diameter/mm | 1.8 | 1.85 |
| Air gap length/mm | 1 | 1 |
| Mass/kg | 16.7 | 16.2 |
| Energy consumption under UDDS cycle/kW. h | 0.0173 | 0.0152 |

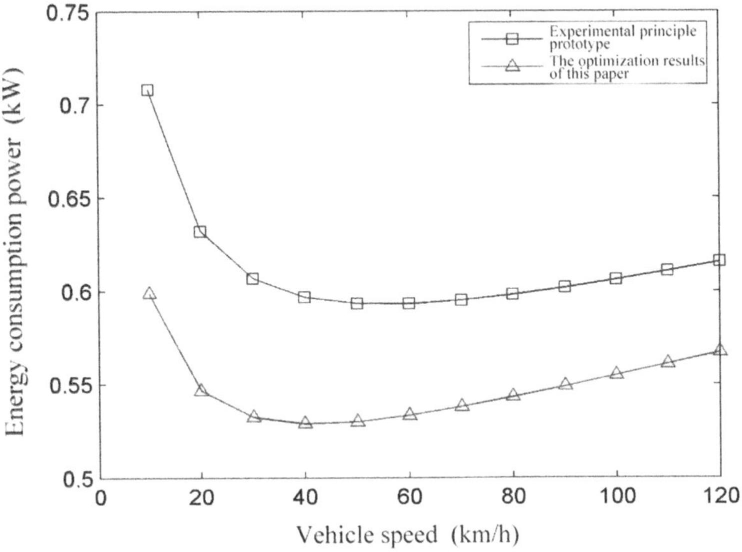


**Fig. 4.18** Contrast curves of the energy-consumption power after optimization

both the center radius of the brake disc and the outer diameter of the coil frame increase and tend to the limit of the constraint conditions. The wire diameter and core diameter of the copper conductor tend to increase, but the increase is not apparent. The energy-saving optimization results show that the total mass of the electromagnetic brake decreases, but the mass decreases little, which is relevant to the weighting coefficient of the mass term in the objective function. The energy consumption based on the UDDS cycle is also investigated in Table 4.1. The results show that the energy consumption based on the UDDS cycle is less than that of the test principle prototype.

As shown in Fig. 4.18, the comparison curve of energy consumption power shows that the energy consumption curve of energy-saving optimization results under different vehicle speeds is better than that of the test principle prototype, which also verifies that the energy consumption of the optimization results based on UDDS cycle conditions is lower than that of the test principle prototype. The "low energy consumption power area" of the energy-saving optimization results shifts to the left, which varies from 30 km/h–80 km/h to 20 km/h–60 km/h compared with the principle prototype; At the same time, as Fig. 4.19 shows, the peak braking torque curve of energy-saving optimization results is only increased by 5 N. M compared with the principle prototype, and the peak braking torque does not increase significantly, but the critical speed of the braking torque curve changes from 500 r/min to about 350 r/min.

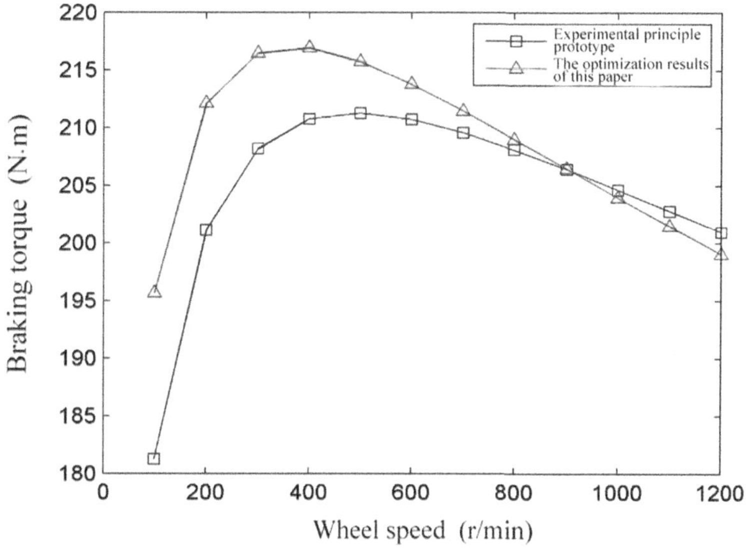


**Fig. 4.19** Contrast curves of the braking torque after optimization

# Chapter 5
# Braking Force Distribution Control of Hybrid Brake-By-Wire System

The braking force distribution ratio of the front and rear axles will affect the directional stability and the utilization of adhesion conditions when the vehicle is driving. For automobile braking systems, it is a very important problem that needs to be handled properly. In this chapter, the braking force distribution of composite brake system for motor and friction and that of integrated brake system for electromagnetic and friction is analyzed and calculated.

## 5.1 Braking Force Distribution Control of Composite Brake System for Motor and Friction

### *5.1.1 Analysis of Influencing Factors on Braking Force Distribution Control*

Figure 5.1 shows the force of the car during braking. In the normal braking state, when the braking intensity of the driver is z, the front axle receives the normal force on the ground:

$$F_{Z1} = \frac{G(b + zh_g)}{L} \tag{5.1}$$

where,

*G*—Car gravity, N;

*b*—Distance from center of mass to rear axle, m;

*hg*—Centroid height, m;

*L*—Car wheelbase, m

The normal ground force on the rear axle:



D. Hu, *Design and Control of Hybrid Brake-by-Wire System for Autonomous Vehicle*, Key Technologies on New Energy Vehicles,
https://doi.org/10.1007/978-981-16-8946-8_5

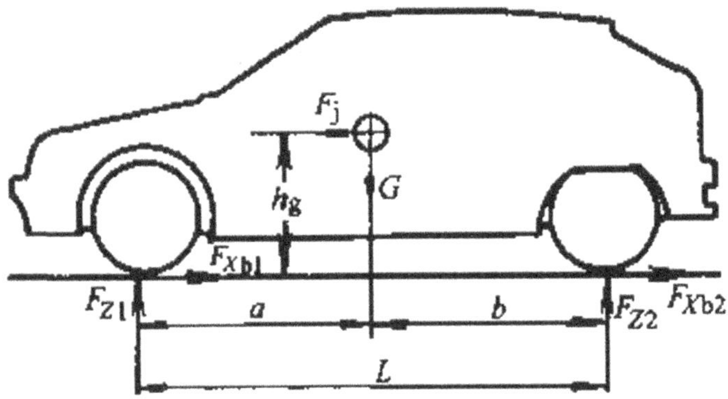


**Fig. 5.1** Force diagram of automotive brake

$$F_{Z2} = \frac{G(a - zh_g)}{L} \tag{5.2}$$

where,

$\alpha$—Distance from center of mass to rear axle, m;

The maximum ground-braking force that the ground can provide to the front axle is obtained through the "road state recognition system" as the pavement adhesion coefficient φ:

$$F_{xb1\max} = F_{Z1}\phi \tag{5.3}$$

The maximum ground braking force that the ground can provide to the rear axle

$$F_{xb2\max} = F_{Z2}\phi \tag{5.4}$$

As Fig. 5.2 [1] shows, the motor torque $T$ as a function of speed $n$ is:

$$T = \begin{cases} T_N & (n < n_0) \\ \frac{9549P_N}{n} & (n > n_0) \end{cases} \tag{5.5}$$

where,

$P_N$—rated power of the motor, KW;

$T_N$—rated torque of the motor, N.m;

$n_0$—rated speed of the motor, r/min;

$n$—motor speed, r/min;

The braking force $F_m$ generated by the motor during the corresponding braking is:

$$F_m = \frac{Ti_g i_0 \eta_T}{r} \tag{5.6}$$

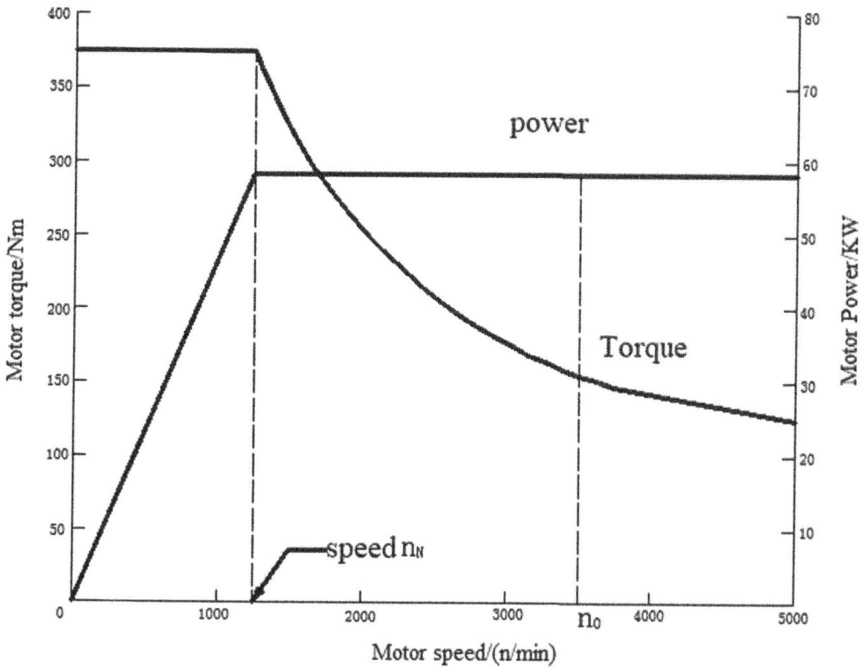


**Fig. 5.2** Typical operating characteristics of the traction motor

where,

$i_g$—the transmission ratio of the transmission;
$i_0$—transmission ratio of the main reducer;
$\eta_T$—transmission efficiency of the transmission system;u
$r$—rolling wheel radius, m;
Motor speed and vehicle speed have the following relationship:

$$u_a = 0.377\frac{rn}{i_g i_0} \tag{5.7}$$

where,

$u\alpha$—Vehicle speed, $km/h$;

When the car starts to decelerate at the initial speed $u_0$ and the braking intensity required by the driver is $z$, then the total braking force what the car needs is $F_{total}$:

$$F_{total} = Gz = F_m + F_{\mu 1} + F_{\mu 2} \tag{5.8}$$

where,

$Fm$—the braking force of the motor, N;
$F\mu 1$—the braking force of the front axle brake, N;
$F\mu 2$—the braking force of the rear axle brake, N。

At this time, the initial value of the angular velocity of the driving wheel $\omega_O = u_O/r$, and the instantaneous angular velocity ω(t) is:

$$\omega(t) = \omega_0 - at \tag{5.9}$$

where,

$\alpha$—Angular acceleration of driving wheel, rad/s2;

$$a = \frac{T}{J} \tag{5.10}$$

where,

$J$—Moment of inertia of driving wheel, kg.m2;

$T$—Braking torque on the driving wheel, N.m;

As shown in Fig. 5.2, the time required for the motor speed to decelerate from $n_0$ to the rated speed $n_N$ is $t_N$:

$$t_N = \frac{\omega_0 - \frac{2\pi n_N}{60 i_g i_0}}{a} \tag{5.11}$$

Figure 5.2 shows the typical operating characteristics of the motor. The regenerative braking of the motor is divided into a constant power interval and a variable power interval. In the variable power interval, the regenerative braking power of the motor becomes smaller. Each time the car brakes for a certain time, the longer the time tN the motor generates in the constant power range, the greater the braking energy recovered by the motor's regenerative brake. As can be seen from the above analysis, the time tN for the motor to generate electricity in the constant power interval is inversely proportional to the braking torque T applied to the drive shaft. To fully reclaim the braking energy of the motor, the frictional braking force applied on the drive shaft should be minimized. For cars equipped with an electro-hydraulic brake system, the above process is achievable. The braking process becomes fully controllable through electro-hydraulic brake. The braking force applied to each wheel can be independently and accurately controlled, and the maximum ground braking force that the ground can apply to the front and rear axles can also be judged before braking.

### 5.1.2 Design and Simulation of Braking Force Distribution Control

To make sure the directional stability of the car and adequate braking efficiency during braking, the ECE R13 braking regulations formulated by the United Nations Economic Commission for Europe put forward clear requirements to the braking

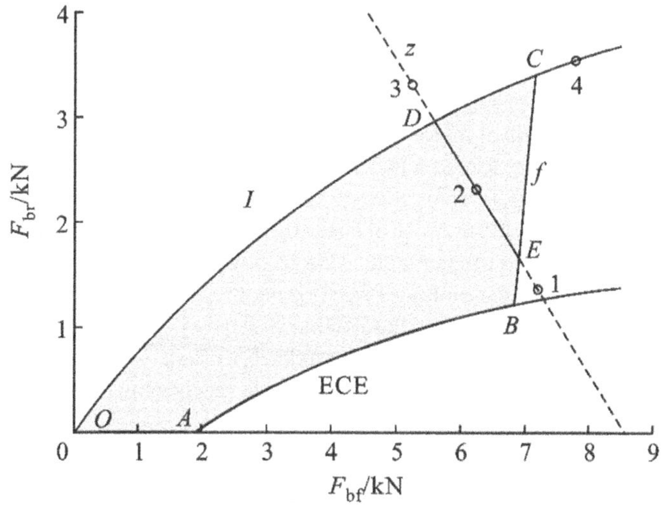


**Fig. 5.3** Schematic diagram of distribution of braking force on the front and rear wheels

force of the front and rear brakes of two-axle vehicles. My country's industry-standard ZBT 24007-1989 also put forward analogous requirements. For cars, the regulations stipulate: For various vehicles between $\psi = 0.2 \sim 0.8$, the braking strength $z \geq 0.1 + 0.85\ (\psi\text{-}0.2)$ is required when the vehicle is in various loading conditions, the front axle should be used with the adhesion coefficient curve. Use the upper part of the adhesion coefficient curve on the rear axle. The traditional braking energy recovery control strategy requires that under the conditions of meeting ECE braking regulations, the design and control follow that the total braking force is distributed to the front wheels as much as possible, that is to say, the distribution of the total braking force will follow the ECE braking. The maximum braking force curve (minimum rear braking force) specified by the regulations is the o-a-b-c curve shown in Fig. 5.3 [2].

As can be seen from the above analysis, to make sure the directional stability of the car and sufficient braking efficiency, the ECE braking regulations put forward two restrictions on the distribution of car braking force: 1. The minimum rear braking force is specified, thereby limiting the braking time. The independent use of the motor on the front axle; Second, it is stipulated that the front axle adhesion coefficient curve should be higher than the rear axle adhesion coefficient curve, which limits the use of the braking energy recovery control strategy described in this article. The ECE regulations set requirements for ordinary braking systems to ensure the safety of braking, but we can see that the invention of electro-hydraulic brake breaks these two limitations. At this time, because electro-hydraulic brake can freely distribute the braking force of the front and rear axles in accordance with the braking intensity and road adhesion coefficient required by the driver, the braking safety of the car can be guaranteed when the entire process of car braking is controlled.

In normal braking mode, when the driver needs the braking strength of z:

(1) If the motor can provide braking power, the $F_m > Gz$ braking power of the whole vehicle is provided entirely by the electric mechanism power, and the braking of the control motor is maintained $F_m = Gz$
    Friction braking force on the front axle $F_{\mu 1} = 0$.
    Friction braking force on the rear axle $F_{\mu 2} = 0$.
(2) If $F_m < Gz$, under the premise of ensuring braking safety, the frictional braking force is allocated to the rear axle as far as possible
a. The rear axle Friction braking force does not reach the threshold.
    Friction braking force on the front axle $F_{\mu 1} = 0$.
    Friction braking force on the rear axle $F_{\mu 2} = Gz - F_m$.
b. The frictional braking power of the rear axle reaches the threshold.
    Friction braking force on the front axle $F_{\mu 1} = Gz - F_m - F_{\mu 2}$.

The threshold of friction braking power on the rear axle $F_{\mu 2\max} = F_{xb2\max} \times s$ is the safety factor. The value of s still needs to be studied to consider the wear of automobile tires and the economy of electric vehicles.

(3) When the car enters the emergency braking state, the "Pavement state recognition system" needs real-time monitoring of the ground attachment $\varphi$ coefficient, the ground attachment coefficient $\varphi$ is passed to the electronic control unit, according to the built-in I curve of the electronic control unit, an ideal vehicle front and rear-wheel braking force distribution is obtained. Then the electronic control unit controls the braking moment of each wheel to meet the requirements. Of course, the specific control process needs further study.

Two evaluation indicators, braking energy recovery rate and vehicle effective energy recovery rate, are used to determine the ability of the control strategy to recover braking energy. The braking energy recovery rate is the ratio of the energy recovered by the energy storage device to the energy consumed by braking, and the effective energy recovery rate of the vehicle is the ratio of the energy recovered by the energy storage device to the energy consumed by the vehicle (driving energy consumption + braking energy consumption).

The braking control strategy in ADVISOR2002 is to look up the meter to determine the electric power coefficient and the front-wheel friction braking force coefficient according to the detected vehicle speed, to realize the braking force distribution control. In the total braking force, assuming that the electric mechanism power coefficient is re and the front-wheel friction braking force coefficient is ff, then the electric mechanism power = total braking force × re, front-wheel friction braking force = total braking force × ff, rear-wheel friction braking force Power = total braking force × (1 – re – ff), the higher the vehicle speed, the greater the electric power coefficient re, and the smaller the front-wheel friction braking coefficient ff. This control strategy has great limitations and does not make full use of electricity. The motivation of the mechanism and whether it can meet the constraints is also unknown.

At present, the widely used braking energy recovery control strategy is the parallel braking force distribution control strategy [3]. The parallel braking control strategy only needs to control the electric mechanism power, and the mechanical braking force is distributed on the basis of the original law, so there are few control parameters and reliability, suitable for the current development of electric vehicles. But this control strategy still takes ECE braking rules as the main boundary condition. For the selection of the switching thresholds z1 and z2 of the parallel braking control strategy, z1 and z2 adopt empirical values of 0.1 and 0.7: when the braking intensity of the car is $z < z1$, only the power of the electric mechanism works; $z1 < z < z2$ When $z > z2$, it is considered as emergency braking, and only the friction braking system works.

Select NYCC operating conditions for simulation. The drive train parameters of the car model are: main reduction ratio $i0 = 1$, $ig = 6.67$, transmission efficiency $\eta T = 0.9$; motor parameters: rated power 75 kW, rated torque 271 N·m, base speed It is 2641 r/min. The simulation results are shown in Table 5.1.

The driving energy consumption and braking energy consumption of the vehicle are related to the selected working conditions and vehicle parameters. It can be seen from Table 1 that the parallel brake distribution control strategy and the brake energy recovery control strategy based on electro-hydraulic compound braking are obvious. It is better than the original braking force distribution control strategy of ADVISOR 2002; at the same time, the braking energy recovery control strategy based on electro-hydraulic compound braking is better than the parallel braking distribution control strategy, and the braking energy recovery of the whole vehicle has increased.

**Table 5.1** Brake energy recovery contrast of different control strategies on NYCC conditions

| | ADVISOR2002 original braking force distribution control strategy | Parallel brake distribution control strategy | Braking energy recovery control strategy based on electro-hydraulic compound braking |
|---|---|---|---|
| Energy storage device recovers braking energy (KJ) | 154 | 302 | 368 |
| Energy consumption of vehicle drive (KJ) | 1273.8 | 1273.8 | 1273.8 |
| Energy consumption of vehicle braking (KJ) | 666.6 | 666.6 | 666.6 |
| Braking energy recovery rate (%) | 23.1 | 45.3 | 55.2 |
| Vehicle effective energy recovery rate (%) | 12.09 | 23.7 | 28.89 |

## 5.2 Braking Force Distribution Control of Integrated Brake System for Electromagnetic and Friction

### *5.2.1 Braking Force Distribution Control in Friction Braking Mode*

Compared with electronic mechanical brakes (EMBs), electronic hydraulic brakes have a backup braking circuit, which can provide a certain braking ability for the vehicle in the case of failure of its energy supply device. However, the realization of the EHB function depends on the performance of the main components, including the EHB controller (electronic circuit), pressure regulating unit (high-speed solenoid valve, motor pump), sensor (pressure sensor, displacement sensor) and so on. Some of these main components are high-precision components, while some have strict requirements on the working environment. Damage to any of these components may cause the failure of the electronic hydraulic brake. Therefore, it is necessary to analyse the safety characteristics of electronic hydraulic brakes.

#### 5.2.1.1 Safety Analysis of Electro-Hydraulic Brake

According to the statistical analysis by the United States Highway Safety Administration, for anti-lock braking systems, the motor pump and high-speed solenoid valve only work 1.4 times over 10,000 braking cycles, but at the beginning of the anti-lock braking system design, the motor pump and high-speed solenoid valve are still required to have a high fatigue life to ensure reliability. However, the motor pump and high-speed solenoid valve in electrohydraulic brake need to complete frequent ordinary braking as well as ensure high safety performance in emergency braking conditions. To prevent the potential risks of EHB and to improve the safety redundancy, improving the fatigue life and reliability of motor pumps and high-speed solenoid valves is necessary and will greatly increase the cost and hinder the wide use of EHBs.

Frequency of Motor Pump Action

When a car is driven in the city, brake use is very frequent. When the electronic hydraulic brake is working, the accumulator needs to continuously provide brake fluid to the friction brake. When the gas pressure of the accumulator is lower than the specified lower limit, the motor pump starts to work to pressurize the accumulator. Therefore, it is necessary to study the factors affecting the action frequency of motor pumps because an action frequency that is too high for the motor pump will lead to fatigue limits and premature failure. According to the analysis of Eq. (4.14), the braking pressure (or braking torque) of the friction brake directly affects the volume

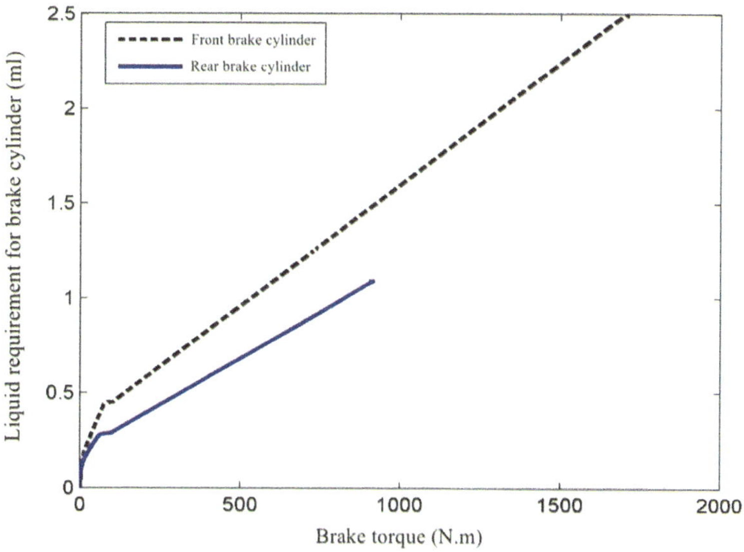


**Fig. 5.4** Relation curves of the brake fluid requirement with the braking torque for the front and rear wheels

of the required brake fluid. As shown in Fig. 5.4, due to the limitation of the design parameters of the brake wheel cylinder piston diameter and brake disc diameter, the output braking torque of the rear friction brake is smaller than that of the front friction brake, but under the condition of the same output braking torque, the volume of brake fluid consumed by the front-wheel brake is larger than that of the rear-wheel brake.

The action frequency of the motor pump can be calculated by the following formula:

$$c = \frac{\sum_{j=1,2,3}^{k} 2V_{fj} + 2V_{rj}}{\Delta V} \tag{5.12}$$

where $V_{fj}$ and $V_{rj}$ are the total fluid demand of the front and rear wheel brakes, respectively, while braking in $m^3$. $j$ is the braking time, and $k$ is the total braking time.

According to formula (5.12), a greater total fluid demand and smaller accumulator displacement result in a higher motor pump action frequency. Therefore, it can be concluded that under the condition that the structural parameters of the main components of the electronic hydraulic brake are determined, reducing the total liquid demand during braking is the main means to reduce the frequency of motor pump action.

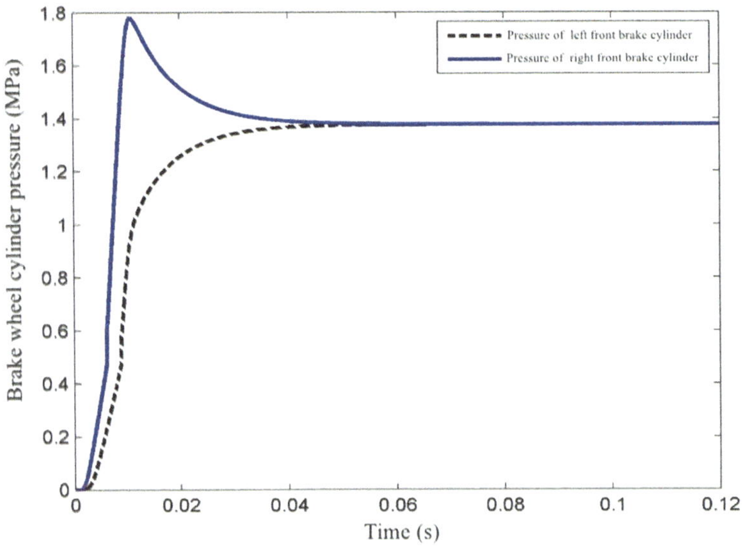


**Fig. 5.5** Pressure curves for the front axle at low braking intensity when there is an inlet valve only on the right wheel's brake

Frequency of High-Speed Solenoid Valve

Under normal braking conditions, the inlet/outlet valve of a single wheel can control the braking force of the left and right friction brakes at the same time due to the existence of a balance valve, for example. The following inspection of three different braking strengths relies only on the front axle right friction brake inlet valve to implement the braking situation. As Fig. 5.5 shows, when the strength of braking is 0.1, the left brake wheel cylinder pressure increases rapidly to the target pressure, while the right brake wheel cylinder pressure has a peak at 0.01 s, and then, the pressure drops to the target pressure. As Fig. 5.6 shows, when the strength of braking is 0.3, the pressure variation characteristics of the left and right brake wheel cylinders are roughly the same as when the braking strength is 0.1, but the pressure wave crest of the right brake wheel cylinder is relatively gentle. The reason for this is that the duty cycle of the inlet valve is relatively small during low-intensity braking, and the inlet valve is closed immediately after being opened for a short time, resulting in the pressure fluctuation of the right brake wheel cylinder.

Comparing Figs. 5.5, 5.6 and 5.7 show that this peak almost disappears when the braking intensity is 0.7, and the increase in the pressure curves for the left and right brake wheel cylinders almost coincide, which verifies the rational of the above explanation. Although there is a certain overshoot of the pressure in the right brake wheel cylinder at low strength, the overshoot time is very short, and considering the elastic deformation of the tire and other factors, it can be completely ignored, while at high strength braking, although the pressures of the left and right brake wheel

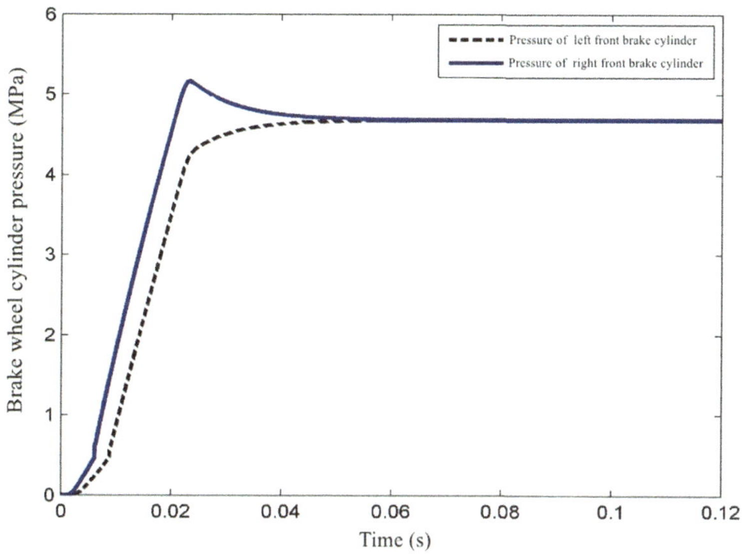


**Fig. 5.6** Pressure curves for the front axle at medium braking intensity when there is an inlet valve only on the right wheel's brake

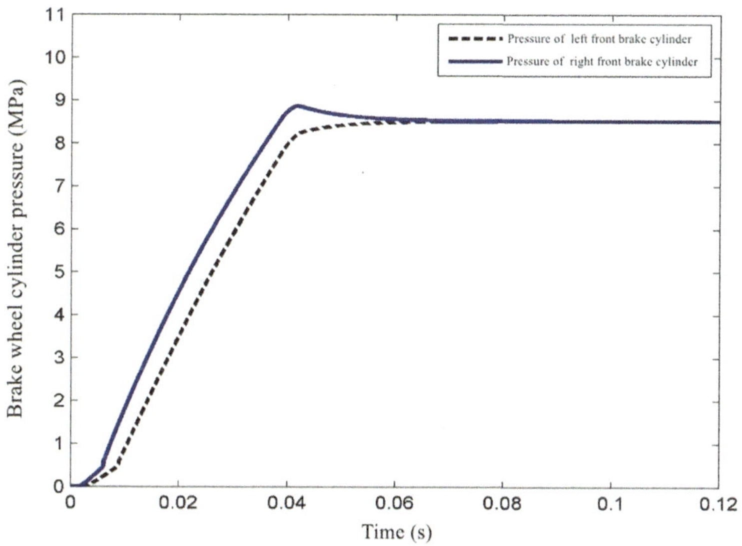


**Fig. 5.7** Pressure curves for the front axle at high braking intensity when there is an inlet valve only on the right wheel's brake

cylinders are almost the same, the vehicle is generally in the emergency braking state, and the braking response is required to be as fast as possible At this time, it is not suitable to only use a single-wheel inlet/outlet valve as the control approach.

#### 5.2.1.2 Braking Force Distribution in Friction Braking Mode

The relationship between the brake fluid demand and brake strength using only the front axle brake and ideal brake force distribution is investigated, as Fig. 5.8 shows. When the brake strength is lower than 0.5, the brake fluid demand based on the ideal brake force distribution is greater than that based on only front axle braking; when the strength of braking is higher than 0.5, the situation is the opposite. This phenomenon seems to conflict with the conclusion in Fig. 5.4. In fact, Fig. 5.4 only shows that the rear-wheel brake alone needs less brake fluid volume than the front brake alone, while Fig. 5.8 shows that when the front and rear axle brakes are used together, the brake fluid required is greater than that of the friction brake alone. The reason for the result shown in Fig. 5.8 is that the volume of brake fluid consumed to eliminate the brake clearance during the braking empty stroke stage of the friction brake is very considerable. Reducing the number of friction brakes participating in braking in a single braking can reduce the volume of brake fluid provided by the accumulator. According to the research results in Sect. 5.1.1, the reduction of the total fluid demand during braking helps to reduce the frequency of motor pump action; therefore, when the demand braking intensity is low, only the braking force should be distributed to

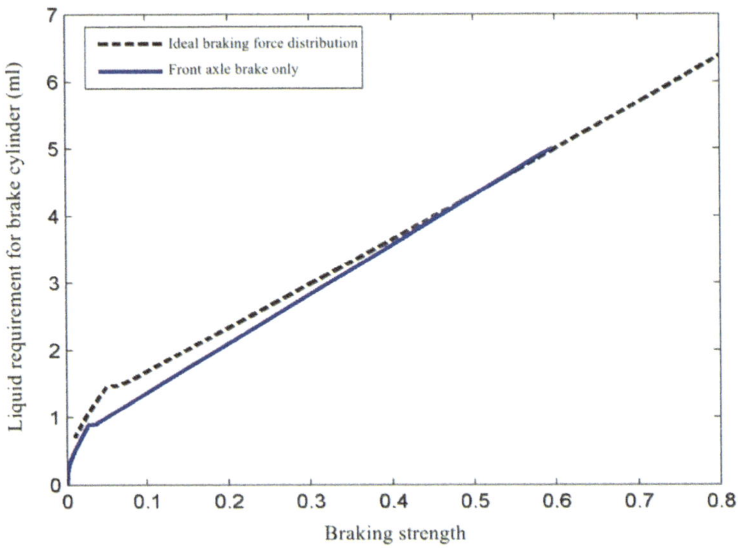


**Fig. 5.8** Relation curves of the brake fluid requirement with the braking force for two braking force distribution methods

**Table 5.2** Key parameters of two typical urban driving cycles

| | ECE-EUDC cycle condition | UDDS cycle condition |
|---|---|---|
| Running time (s) | 1225 | 598 |
| Mileage (km) | 10.93 | 1.9 |
| Maximum speed (km/h) | 120 | 44.58 |
| Average speed (km/h) | 32.1 | 11.41 |
| Maximum braking deceleration ($m/s^2$) | −1.39 | −2.64 |
| Average braking deceleration ($m/s^2$) | −0.79 | −0.61 |
| Idle time (s) | 339 | 210 |
| Number of stops | 13 | 18 |

the front axle; when the demand braking intensity is high, the braking force of the front and rear axles should be distributed on the basis of the I curve to reduce the frequency of motor pump action.

At the same time, the need to reduce the frequency of high-speed solenoid valves must be given enough attention. The influence of typical urban vehicle driving conditions on the front and rear axle braking force distribution method is analysed, and the ECE-EUDC and UDDS cycles are selected for analysis. As shown in Table 5.2, most of the braking decelerations in the two typical urban vehicle driving conditions are less than 1 $m/s^2$, and the low-intensity braking demand is quite frequent. Therefore, when the braking strength is less than 0.1, the braking force of the car is fully distributed to the front axle so that the rear axle does not brake, and the frequency of the rear axle high-speed solenoid valve use can also be reduced (Table 5.2).

Considering the need to reduce the frequency of the motor pump and high-speed solenoid valve based on the ECE R13 braking regulation requirements, the braking force distribution method of the front and rear axles of the friction and electromagnetic integrated braking system when using friction braking mode is preliminarily proposed: when the strength of braking is lower than 0.1, only front axle braking is used; when the strength of braking is higher than 0.1, the braking forces of the front and rear axles are distributed on the basis of the I curve, as Fig. 5.9 shows.

Simultaneously, the research results in Sect. 5.1.1 provide a theoretical basis for further reducing the frequency of the high-speed solenoid valves. Therefore, in friction braking mode, the front and rear axle braking force distribution method using the integrated friction and electromagnetic braking system should be as follows: when the strength of braking is less than 0.1, the front axle balance valve is opened, and the inlet/outlet valves of the left and right wheels on the front axle control the two friction brakes of the front axle in turn, as shown in Fig. 5.10 (BOLD); when the strength of braking is higher than 0.1 and less than 0.7, the front and rear axle balance valves are opened, and the inlet/outlet valves of the left and right wheels on the front and rear axles control the braking of the two front axle friction brakes in turn, as

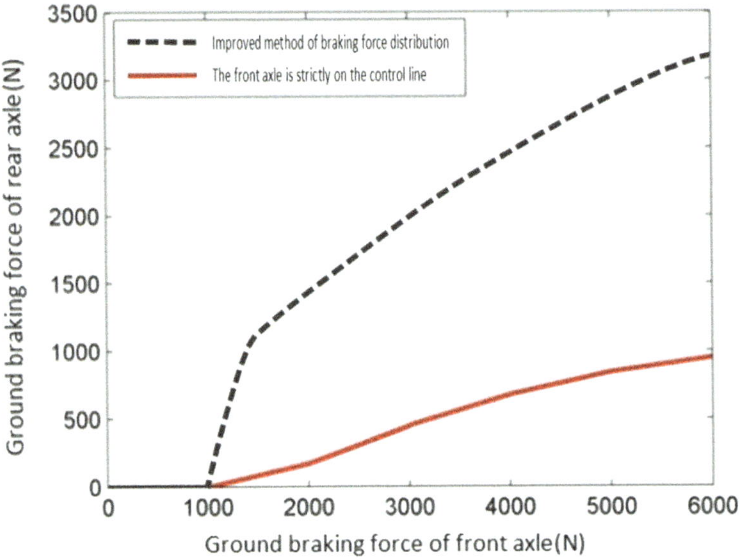


**Fig. 5.9** Curves of the distribution method of the front and rear braking force distributions when using friction braking mode

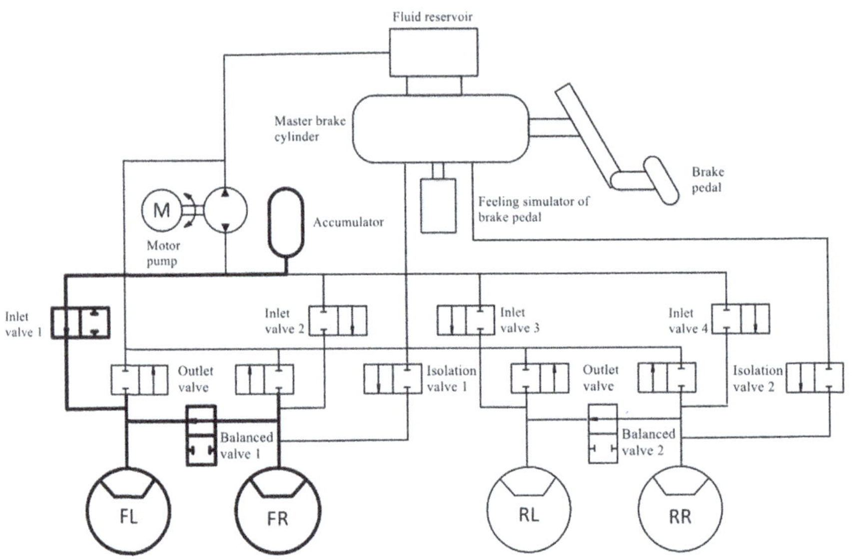


**Fig. 5.10** EHB working condition when only the inlet valve of the left wheel in the front axle is used

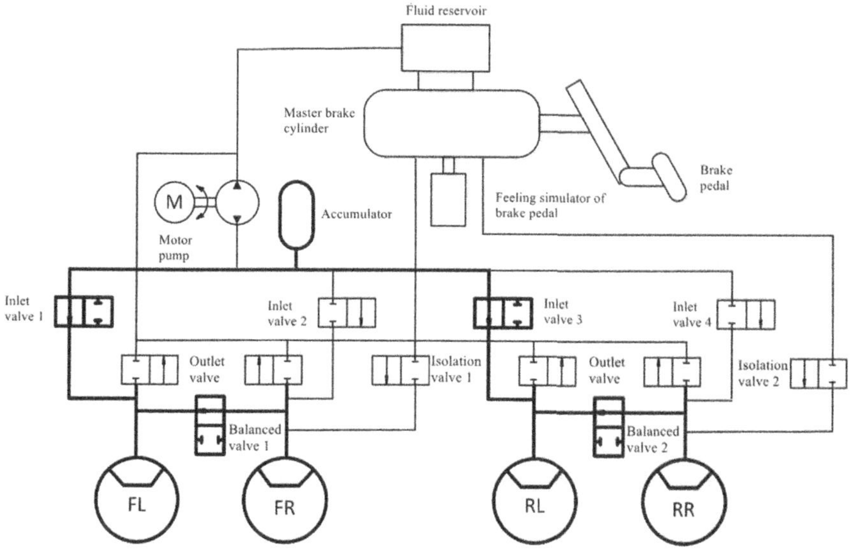


**Fig. 5.11** EHB working condition when the inlet valve of the left wheel in both the front and rear axles are used

shown in Fig. 5.11 (BOLD). When the strength of braking is higher than 0.7, the vehicle enters the emergency braking state, the front and rear axle balancing valves are closed, and each wheel is controlled by its own corresponding inlet/outlet valve.

Finally, the ideal braking force distribution method and the improved braking force distribution method (i.e., the front and rear axle braking force distribution method when using friction braking mode) are simulated. After 10 UDDS cycles, the influence of the two braking force distribution methods on electrohydraulic braking's safety is shown in Table 5.3. The number of motor pump actions is reduced by 43%, the number of front axle inlet valve actions is reduced by 50%, and the number of rear axle inlet valve actions is reduced to 8.6% in view of the ideal braking force distribution. This shows that compared with the ideal braking force distribution method, the improved braking force distribution method can effectively reduce the use frequency of the motor pump and the high-speed solenoid valves, prevent premature fatigue damage, and ensure the safety of electrohydraulic brake.

**Table 5.3** Comparison of the EHB security features with different braking force distribution methods over ten UDDS driving cycles

| | Motor pump action times | Front axle inlet valve action times | Rear axle solenoid valve action times |
|---|---|---|---|
| Ideal braking force distribution method | 7 | 350 | 350 |
| Improved braking force distribution method | 4 | 175 | 30 |

### 5.2.2 Braking Force Distribution Control in Electromagnetic Braking Mode

During the continuous braking process of the electromagnetic brake, the resistance value of the excitation coil will increase due to the increase in the temperature. According to formulas (4.15)–(4.17), when the demand excitation current is constant, additional power consumption will be generated; when the brake disc temperature rises, the electromagnetic braking torque will produce a certain degree of attenuation, and the output electromagnetic braking torque will be constant. In this circumstance, the electromagnetic brake will also need to consume more energy, so it is necessary to study the influence of the thermodynamic characteristics of the electromagnetic brake on its energy consumption.

#### 5.2.2.1 Thermodynamic Modelling of Electromagnetic Brake

In the analysis of the thermodynamic characteristics of the electromagnetic brake, the following assumptions are made: (1) the air temperature around the electromagnetic brake is the same as the ambient temperature; (2) there is no relative slip between the wheel and the road during the operation of the electromagnetic brake; (3) the heat conduction of the electromagnetic brake is only carried out inside the components during the continuous operation of the electromagnetic brake; and (4) due to the strong heat conduction ability of the brake disc, it is recognized that the temperature distribution on the brake disc is uniform.

When the electromagnetic brake works, the kinetic energy of the vehicle is converted into heat energy to be dissipated. Part of the heat energy is stored in the brake disc, which makes the temperature of the brake disc increase continuously; the other part of the heat energy is emitted into the air by the continuously rotating brake disc [4, 5]. On the basis of the law of conservation of energy, the kinetic energy absorbed by the electromagnetic brake in unit time is equal to the sum of the heat energy emitted by the electromagnetic brake and the heat energy stored in unit time. Therefore, according to the law of conservation of energy, it can be concluded that

$$Q_d = P_{eb} - P_d \tag{5.13}$$

where $Q_d$ is the heat energy stored in brake disc per unit time in kw, $P_{eb}$ is the heat energy absorbed by the electromagnetic brake in unit time in kw, and $P_d$ is the heat energy emitted by the brake disc per unit time in kw.

The heat energy emitted by the brake disc in unit time is calculated by the following formula:

$$P_d = h_d A_d (T_{d1} - T_{d0}) \tag{5.14}$$

where $h_d$ is the sum of convective and radiative heat transfer coefficients of the brake disc, $A_d$ is the effective heat dissipation area of the brake disc in m$^2$, and $T_{d0}$ and $T_{d1}$ are the brake disc temperature before and after braking, respectively, in °C.

The air Reynolds number is defined as the ratio of inertial force and internal friction force. The expression of air Reynolds number is:

$$R_{ea} = \frac{v_a \rho_a}{\mu_a} L_a \tag{5.15}$$

where $v_a$ is the air kinematic viscosity in m$^2$/s, $\rho_a$ is the air density in kg/m$^3$, $\mu_a$ is the air viscosity in Pa.s, and $L_a$ is the characteristic length defined by the Reynolds number in m.

When $R_{ea} > 2.4 \times 10^3$, the convective heat transfer coefficient can be expressed as

$$h_{con} = 0.04\left(\frac{K_{air}}{r_2}\right)R_{ea}^{0.8} \tag{5.16}$$

where $K_{air}$ is the thermal conductivity of air.

When $R_{ea} \leq 2.4 \times 10^3$, the convective heat transfer coefficient can be expressed as

$$h_{con} = 0.7\left(\frac{K_{air}}{r_2}\right)R_{ea}^{0.55} \tag{5.17}$$

The radiant heat transfer coefficient of the brake disc is determined by the following formula:

$$h_{rad} = \frac{\theta\zeta(T_{d1}^4 - T_{d0}^4)}{T_{d1} - T_{d0}} \tag{5.18}$$

where $\zeta$ is the Stefan-Boltzmann constant, and $\theta$ is the emissivity of the brake disc.

The heat stored per unit time of the brake disc can also be calculated by the following formula:

$$Q_d = \frac{m_d c_d \left(T_{d(t+\Delta t)} - T_{dt}\right)}{\Delta t} \tag{5.19}$$

where $c_d$ is the brake disc specific heat in J/(kg·°C), $T_{d(t+\Delta t)}$ is the temperature of the brake disc at t + Δt in °C, and $T_d$ is the temperature of the brake disc at time t in °C.

If Δt in Eq. (5.19) tends to zero, the equation becomes

$$Q_d = m_d c_d \lim_{\Delta t \to 0} \frac{\left(T_{d(t+\Delta t)} - T_{dt}\right)}{\Delta t} = m_d c_d T_d \tag{5.20}$$

where $T_d$ is the instantaneous temperature of the brake disc in °C.

Substituting Eq. (4.16), Eq. (5.14) and Eq. (5.20) into Eq. (5.13), the expression of the instantaneous temperature of the brake disc can be obtained as

$$m_d c_d T_d = \frac{N_p S_p^2 \Delta_h \omega_d^3}{8\pi \rho_d k_a} B^2 - h_d A_d (T_{d1} - T_{d0}) \tag{5.21}$$

The analysis of the instantaneous temperature of the excitation coil is similar to the above analysis and will not be repeated here. According to Eq. (5.21), the thermodynamic model of electromagnetic brake is established to study the braking force distribution method when using electromagnetic braking mode.

#### 5.2.2.2 Braking Force Distribution in Electromagnetic Braking Mode

According to the above analysis, when using an electromagnetic brake for automobile braking, the temperature of the excitation coil and brake disc increases with increasing braking time and braking strength, and the increase in the temperature of the excitation coil and brake disc further improves the resistance and current of the excitation coil, which increases the energy consumption of the electromagnetic brake. In this circumstance, research on the energy-saving control method for electromagnetic braking makes it very meaningful to use the least electric energy when completing the same braking task. There are two problems to be solved when researching an energy-saving control method for electromagnetic brake: the braking task distribution between the excitation coils in the braking process of a single electromagnetic brake and the distribution of the braking torque between the front and rear axle electromagnetic brakes. Therefore, two groups of comparative experiments are designed in this paper.

In the first group of tests, two groups of excitation coils and four groups of excitation coils are used to engage the electromagnetic brake under the condition that the required braking torque is the same, as shown in Figs. 5.12 and 5.13. Curve 1 is the simulation result of braking with all four excitation coil groups, while curve 2 is the simulation result of braking with only two excitation coil groups. In the two braking processes, the electromagnetic braking torque is 100 N.m, and the wheel speed is 400 R/min. Figure 5.13 shows that using all four excitation coil brake groups consumes less power than using only two excitation coil brake groups. Because the braking torque and speed are the same for both test conditions, the braking energy absorbed by the brake disc and the increase in the temperature of the brake disc are almost the same. However, Fig. 5.12 shows that the temperature of all four groups of excitation coils increases slowly and that the amplitude of increase is lower than that of the two groups with only excitation coils. This shows that the main reason for the poor energy consumption is that the temperature of the excitation coil is too high when only two excitation coil groups are used for braking.

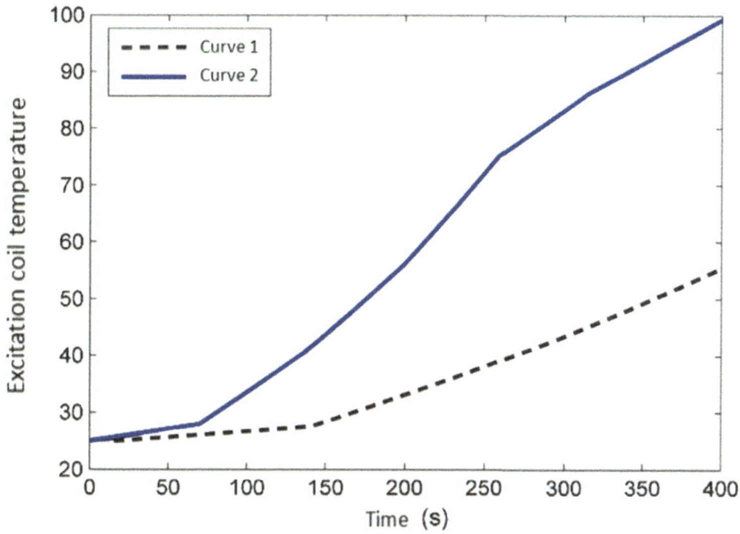


**Fig. 5.12** Temperature curves of the ECB's field winding with different numbers of pole pairs

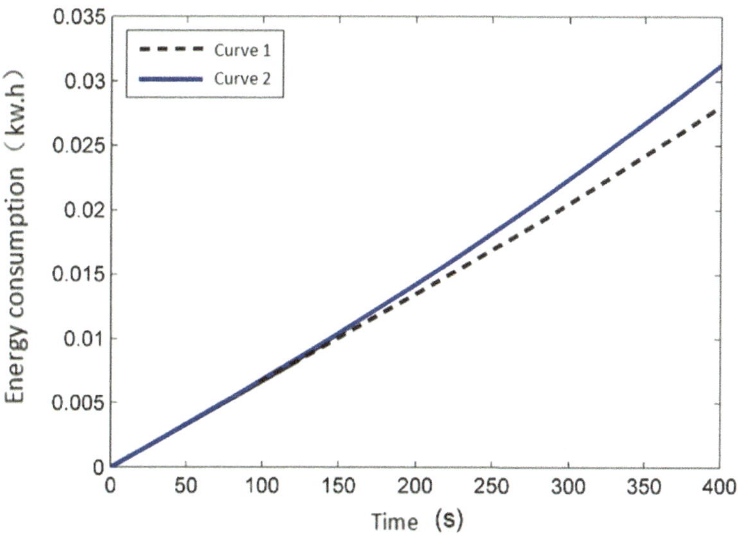


**Fig. 5.13** Energy-consumption curves of the ECB with different numbers of pole pairs

The second group of tests compares the total electric energy consumption for the situation where a 400-N.m braking task is equally distributed to only two electromagnetic brakes on the front axle and where a 400-N.m braking task is equally distributed to four electromagnetic brakes on the front axle and rear axle, as Figs. 5.14, 5.15 and 5.16 shows. The wheel speed remains constant at 400 R/min. Curve 1 shows that all

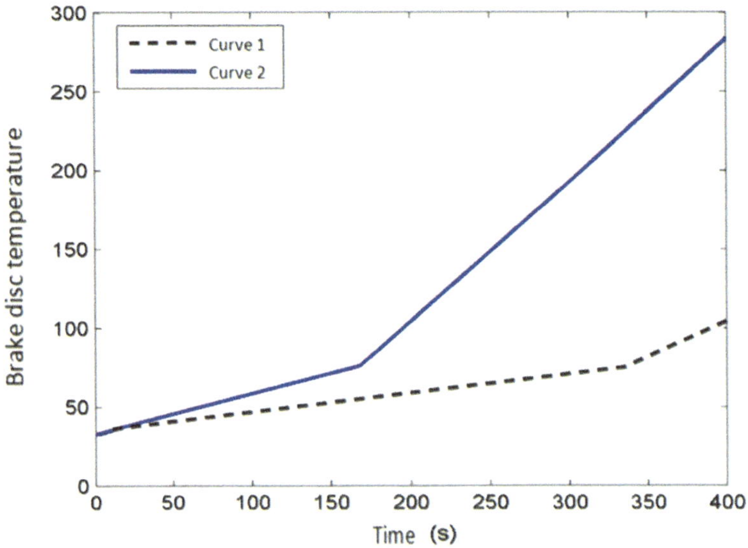


**Fig. 5.14** Temperature curves of the ECB's brake discs for different brake tasks

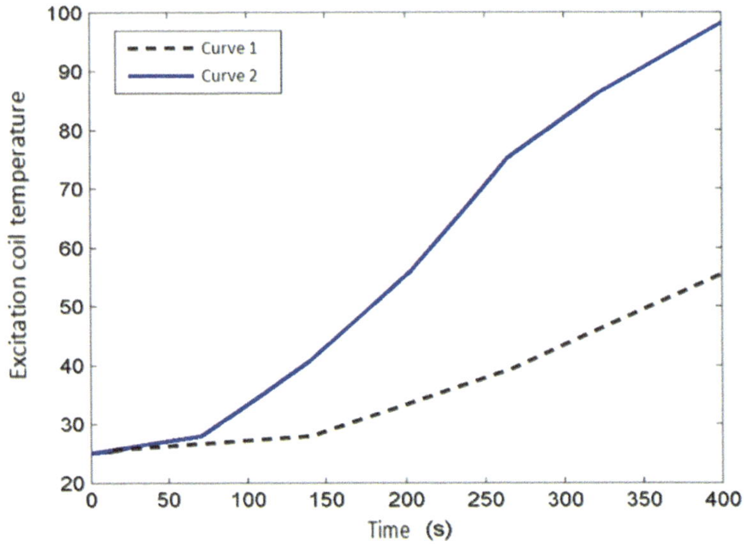


**Fig. 5.15** Temperature curves of the ECB's field winding for different brake tasks

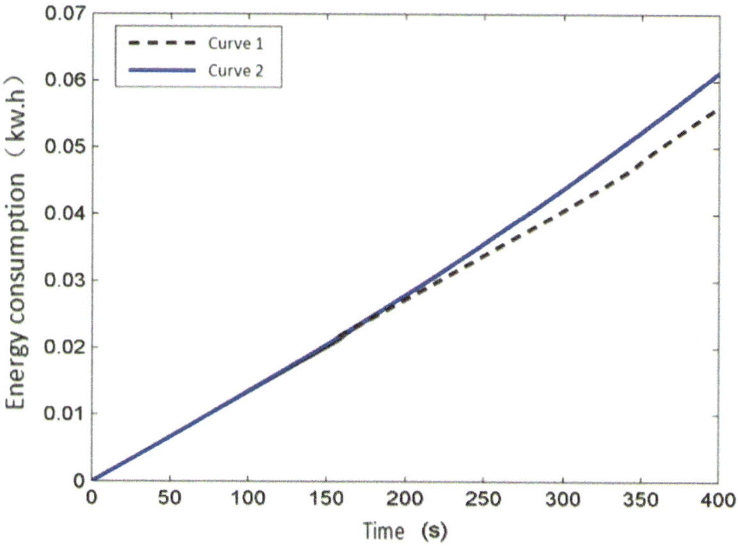


**Fig. 5.16** Energy-consumption curves of the ECB for different brake tasks

four electromagnetic brakes on the front and rear axles work, and the braking torque output by a single electromagnetic brake is 100 N.m; curve 2 shows that only two electromagnetic brakes on the front axle work, and the braking torque output by a single electromagnetic brake is 200 N.m. Obviously, a higher output braking torque results in a higher corresponding temperature, regardless of whether it is the excitation coil temperature or the brake disc temperature, as Figs. 5.1 and 5.12 shows. Distributing the braking force to one axle consumes more energy than distributing the braking force to the front and rear axles equally. The reason for the excessive electric energy consumption is that the temperatures of the excitation coil and brake disc are too high, as shown in Fig. 5.13.

In summary, when using an electromagnetic brake, the braking task is evenly distributed, which can prevent the increase in the temperature of a single excitation coil or brake disc from being too high and achieve the purpose of reducing energy consumption. The corresponding energy-saving control method of an electromagnetic brake is as follows: when using a single electromagnetic brake, all the excitation coils should be controlled to work at the same time so that the excitation current is the same. When using the electromagnetic brake mode and distributing the braking force of the front and rear axles, considering the limitation of the braking regulations, the electromagnetic braking torque of the front and rear axles should be distributed according to the ideal braking force distribution curve of the front and rear axles.

### 5.2.3 Braking Force Distribution Control in Combined Braking Mode

In the process of automobile braking, it is necessary to investigate the reasonable degree of the braking force distribution method for the front and rear axles. It is generally believed that when the utilization adhesion coefficient is closer to the braking strength, the ultimate adhesion provided by the ground will be fully utilized. Therefore, the utilization adhesion coefficient of the front and rear axles is defined as

$$\varphi_i = \frac{F_{xbi}}{F_{zi}} \tag{5.22}$$

In the equation, $\varphi_i$ is the utilization adhesion coefficient of the i-th axis, $F_{xbi}$ is the ground braking force in N, and $F_z$ is the ground normal reaction in N.

As Fig. 5.17 [1] shows, the ground normal reaction force on the front and rear axles of the vehicle is as follows:

$$\begin{cases} F_{zf} = \dfrac{G}{l}(l_r + zh_g) \\ F_{zr} = \dfrac{G}{l}(l_f - zh_g) \end{cases} \tag{5.23}$$

where $F_{zf}$ and $F_{zr}$ are the normal ground reaction of the front and rear axles, respectively, in N, $l_f$ and $l_r$ are the distance between the centre of mass of the car and the centreline of the front and rear axles, respectively, in m, $G$ is the automobile's gravity in N, $l$ is the wheelbase in m, $z$ is the braking strength, and $h_g$ is the centroid height in m.

The formula for calculating the ground braking force of the front and rear axles is

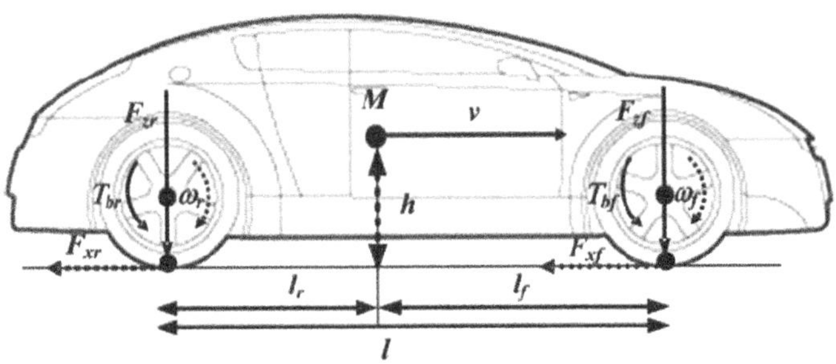


**Fig. 5.17** Force diagram of t vehicles while braking

$$\begin{cases} F_{xbf} = Gz\beta \\ F_{xbr} = Gz(1 - \beta) \end{cases} \tag{5.24}$$

where $F_{xbf}$ and $F_{xbr}$ are the ground braking force of the front and rear axles, respectively, in N, and $\beta$ is the brake force distribution coefficient.

According to Eq. (5.23) and Eq. (5.24), the available adhesion coefficients of the front axle and the rear axle can be obtained:

$$\begin{cases} \varphi_f = \dfrac{l\beta z}{l_r + zh_g} \\ \varphi_r = \dfrac{l(1 - \beta)z}{l_f - zh_g} \end{cases} \tag{5.25}$$

where $\varphi_f$ and $\varphi_r$ are the utilization adhesion coefficients of the front axle and the rear axles, respectively.

In order to make sure the stability of the vehicle braking direction and make full use of the ground adhesion conditions, proposed ECE R13 braking regulations make clear provisions on the distribution of braking force between the front and rear axles of two-axle vehicles: (1) when the vehicle is driving on the road with an adhesion coefficient between 0.2 and 0.8, the braking strength and utilization adhesion coefficient shall meet the following requirements: $\varphi \le (Z + 0.07)/0.85$; (2) the utilization adhesion coefficient of the front axle of the vehicle must be greater than that of the rear axle without regard to the loading state; (3) when the vehicle is running on a road with an adhesion coefficient between 0.2 and 0.8, the utilization adhesion coefficient of the front axle of the vehicle must be greater than that of the rear axle. The braking strength is between 0.3 and 0.4, and the rear axle utilization adhesion coefficient does not exceed the straight line $\varphi = Z + 0.05$. In this case, the rear axle utilization adhesion coefficient can be greater than the front axle utilization adhesion coefficient. According to the ECE R13 requirements, the coefficient of adhesion of the front and rear axles adheres to the following equation:

$$\begin{cases} \varphi_f \le \dfrac{z + 0.07}{0.85} \\ \varphi_r \le \dfrac{z + 0.07}{0.85} \\ \varphi_f > \varphi_r \end{cases} \tag{5.26}$$

By substituting Eq. (5.25) into Eq. (5.26), the braking force distribution coefficient satisfies the following equation:

$$\begin{cases} \beta \leq \dfrac{(z+0.07)(l_r + zh_g)}{0.85zl} \\ \beta \geq 1 - \dfrac{(z+0.07)(l_f - zh_g)}{0.85zl} \\ \beta > \dfrac{l_r + zh_g}{l} \end{cases} \tag{5.27}$$

As Eq. (5.26) shows, the front and rear axle braking force distribution coefficient control curves can be obtained by substituting the vehicle parameters. The specific vehicle model parameters are as follows: the front axle load is 840 kg, the rear axle load is 560 kg, and the centre of mass height is 0.56 m under no load; the front axle load is 960 kg, the rear axle load is 815 kg, and the centre of mass height is 0.544 m under full load; the wheelbase is 2.803 m, and the wheel rolling radius is 0.327 m. The diameter of the brake piston is 60 mm, the diameter of the front brake disc is 280 mm, the diameter of the rear brake piston is 42 mm, and the diameter of the rear brake disc is 225 mm.

When the braking force distribution coefficient is higher than the locking sequence control line of the front and rear axles, the front axle is locked before the rear axle to avoid side slip of the rear axle; when the braking force distribution coefficient is below the strict upper control line of the front axle and above the strict lower control line of the rear axle, the utilization adhesion coefficient of the front axle and the rear axle can meet ECE R13 requirements, as shown in Fig. 5.18. Therefore, according to ECE R13 requirements, the distribution range of the braking force coefficient should

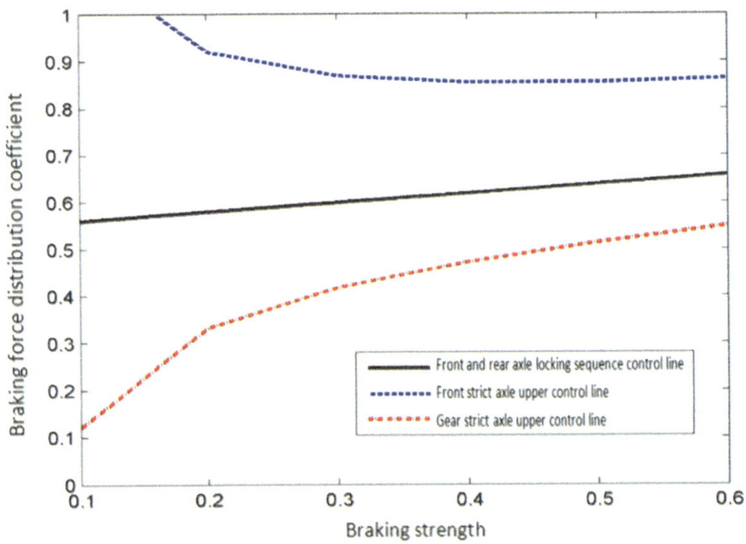


**Fig. 5.18** Relation curves of braking force distribution coefficients and braking intensity based on ECE R13 regulation

be distributed between the front axle upper control line and the front and rear axle locking sequence control line.

The integrated brake system for electromagnetic and friction integrates electro-hydraulic brake and electromagnetic brake on four wheels and can independently control the braking force of the four wheels. Differing from a traditional hydraulic brake system, the front and rear axle braking forces need to be set as a fixed ratio. The combined braking torque output by the friction and electromagnetic integrated braking system can be distributed to the front and rear axles according to the ideal braking force distribution curve. Therefore, when using combined braking mode, the formula for the ground braking force of the front and rear axles is as follows:

$$\begin{cases} F_{xbf} = Gz\dfrac{l_r + zh_g}{l} \\ F_{xbr} = Gz\dfrac{l_f - zh_g}{l} \end{cases} \tag{5.28}$$

According to the above analysis, when using the combined braking mode, the combined braking torque of the front axle or the rear axle of the vehicle has been determined. The braking force of the electro-hydraulic brake and the electromagnetic brake on the front or rear axle should be coordinated according to their respective characteristics and the driver's braking demand. This will be discussed in Chap. 6.

## References

1. Medrdad E, Gao Y, Ali E (2011) Modern electric vehicles, hybrid electric vehicles and fuel cell vehicles—the basic principles, theory and design [M]. Machinery Industry Press, Beijing
2. Zhisheng YU (2000) Automobile theory [M]. Machinery Industry Press, Beijing
3. Hu D, Wang Y, Li J et al (2021) Investigation of optimal operating temperature for the PEMFC and its tracking control for energy saving in vehicle applications[J]. Energy Convers Manag 249:114842
4. Yoshida K, Tasaka M (2004) Convective heat transfer from a rotating drum with fin array for the permanent magnet type eddy current retarder[J]. Therm Sci Eng 12(6): 45–52
5. Liu C, Jiang K, Zhang Y (2010) Study on temperature rise performance of eddy current retarder in automobile[C]

# Chapter 6
# Mode Switching Control of Hybrid Brake-By-Wire System

When the braking system switches between different braking modes, different switching control laws will affect the stability of the system. Therefore, the braking mode switching control method is of great significance to ensure its stability and safety. In this chapter, the system braking mode switching control strategy under emergency braking and non-emergency braking is proposed. The stability in the braking mode switching process is analyzed by using the common Lyapunov function method, and the decoupling controller of integrated brake system for electromagnetic and friction is designed.

## 6.1 Mode Switching Control of Hybrid Brake-By-Wire System Under Non-emergency Braking

### *6.1.1 Stability Analysis of Braking Mode Switching Process*

#### 6.1.1.1 Braking Mode Switching Control Strategy

The braking mode switching control strategy of integrated brake system for electromagnetic and friction needs to judge the braking method of the integrated brake system according to the driver's demand for braking strength change rate, wheel speed, steering wheel speed, etc. yaw rate, and lateral acceleration. The vehicle braking conditions are divided into standard braking and emergency braking. Emergency braking should include active functions such as anti-lock braking control, driving anti-skid power, electronic braking force distribution control, and electronic stability control. Next, the braking mode switching control strategy of the frictional and electromagnetic integrated brake system is proposed as follows.



D. Hu, *Design and Control of Hybrid Brake-by-Wire System for Autonomous Vehicle*, Key Technologies on New Energy Vehicles,
https://doi.org/10.1007/978-981-16-8946-8_6

(1) When the driver depresses the brake pedal, judge whether it is normal braking condition or emergency braking condition on the basis of the driver's required braking intensity and the change rate of braking power, or the integrated brake system controller directly judges whether active braking intervention is necessary according to the vehicle running state.
(2) If the driver carries out standard braking, an integrated brake system for electromagnetic and friction first enters the electromagnetic braking mode. When the driver expects the braking intensity to be lower than 0.1, it will always be in the electromagnetic braking mode.
(3) When the driver required braking intensity exceeds 0.1, the integrated brake system enters the combined braking mode. In the combined braking mode, the electro-hydraulic brake is used to meet the maintenance of braking strength in the driver braking demand, and electromagnetic brake are used to meet the change of braking strength in the drive braking direction to coordinate and control the electro-hydraulic brake and electromagnetic brake.
(4) When the integrated brake system stops the electromagnetic braking when the wheel speed is lower than 150 r/min, and switches from the combined braking mode to the friction braking mode. Similarly, suppose the change rate of braking intensity required by the driver is zero, and the working time of electromagnetic brake exceeds 3S. In that case, the integrated brake system switches from combined braking mode to friction braking mode.
(5) In case of emergency braking, considering that the vehicle wheel speed is generally high, it is proposed to switch the braking mode according to the braking strength requirements of each wheel. When a single wheel's calculated expected braking strength is lower than 0.1, enter the electromagnetic braking mode, otherwise enter the combined braking mode. In the combined braking mode, it is required to coordinate and control the electro-hydraulic brake and electromagnetic brake. Electro-hydraulic brake is mainly used to meet the braking strength demand, and electromagnetic brake is used primarily to hold the wheel state.

#### 6.1.1.2 Hybrid Characteristic Modeling of Hybrid Brake-By-Wire System

Hybrid Petri net is a Petri net mode based on the hybrid idea, which is mainly used for the modeling and analysis of hybrid dynamic systems. Hybrid Petri nets include continuous Petri nets and discrete Petri nets. Continuous Petri nets are used to simulate the dynamic changes of continuous systems; Discrete Petri net is used to simulate discrete system state changes. Its continuous part and discrete part can interact directly at the same level. Figure 6.1 [1, 2] is the schematic diagram of hybrid characteristic modeling of integrated brake system for electromagnetic and friction using hybrid Petri net, in which the hybrid dynamic behavior of integrated brake system for electromagnetic and friction is analyzed as follows:

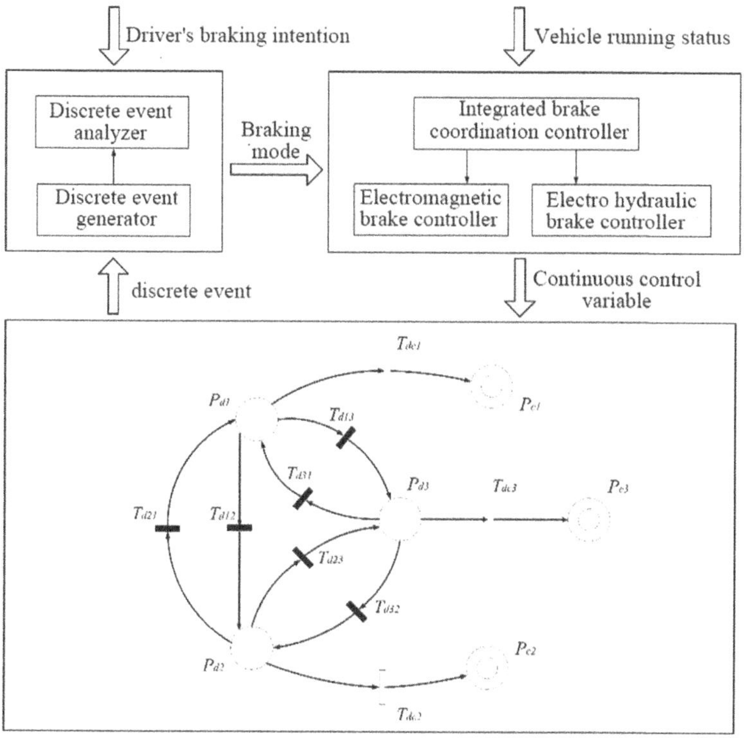


**Fig. 6.1** Hybrid Petri net mode of integrated brake system for electromagnetic and friction

When the driver gives the braking command according to the road conditions, the initial database of hybrid Petri net mode of integrated brake system for electromagnetic and friction starts to input the driving state information of the vehicle and the braking intention of the driver, and the controller of integrated brake system for electromagnetic and friction judges the initial braking conditions. The controller of integrated brake system for electromagnetic and friction determines the discrete transition of integrated brake system for electromagnetic and friction according to the discrete transition occurrence rules and the final discrete warehouse obtained after the transition (including $P_{d1}$, $P_{d2}$, $P_{d3}$, $P_{d1}$ is electromagnetic braking mode, $P_{d2}$ is friction braking mode and $P_{d3}$ is combined braking mode), so as to realize the selection and switching of three braking modes. The occurrence rule of discrete transition ($T_{d12}$, $T_{d21}$, $T_{d23}$, $T_{d32}$, $T_{d13}$, $T_{d31}$) of integrated brake system for electromagnetic and friction is formulated according to the braking mode switching control strategy proposed in Sect. 6.1.1. When the integrated brake system for electromagnetic and friction completes the selection or switching of braking mode, the integrated brake system will carry out corresponding discrete–continuous transition ($T_{dc1}$, $T_{dc2}$, $T_{dc3}$), and the change of continuous vehicle state (I.e., continuous depot, $P_{c1}$, $P_{c2}$, $P_{c3}$) can be realized according to the discrete–continuous transition.

#### 6.1.1.3 Braking Stability Analysis of Hybrid Brake-By-Wire System

Ensuring the stability of integrated brake system for electromagnetic and friction in the process of braking mode switching is the premise of studying the braking mode switching control method. The stability of nonlinear switched systems (special hybrid dynamic systems) composed of continuous nonlinear subsystems and discrete uncertain subsystems, such as integrated brake system for electromagnetic and frictions, is much more complex than that of conventional switched systems composed of continuous or discrete subsystems.

The general nonlinear switched system is considered in the following form:

$$\begin{cases} x = f(x, u, t)x \in R^n, u \in R^m, t \in R \\ y = h(x)x \in R^s, n > m > s \end{cases} \tag{6.1}$$

**Theorem 1** *For nonlinear switched systems, if there is a positive definite Lyapunov function* $V_i(x(t),t)$, *Class K function* $\alpha_i$. i = 1, 2,…, p, *satisfying the following conditions*:

(1) $\alpha_i(\|x(t)\|) \leq V_i(x(t), t), i = 1, 2, \ldots, p$;
(2) $\dot{V}_1(x(t), t) \leq 0, i = 1, 2, \ldots, p, x \in B_r = \{x: \|x\| \leq r, r > 0\}$;
(3) $V_{i+1}(x(t),t) \leq V_i(x(t),t), i = 1, 2, \ldots, p$;

*Then the equilibrium point is stable in the sense of Lyapunov.*

**Theorem 2** *For nonlinear switched systems, if there is a positive definite Lyapunov function* $V_i(x(t),t)$, *Class K function* $\alpha_i$ *and* $\beta_i$, i = 1, 2,…, p, *satisfying the following conditions*:

(1) $\alpha_i(\|x(t)\|) \leq V_i(x(t),t) \leq \beta_i(\|x(t)\|), i = 1, 2, \ldots, p$;
(2) $\dot{V}_1(x(t), t) \leq 0, i = 1, 2, \ldots, p, x \in B_r = \{x: \|x\| \leq r, r > 0\}$;
(3) $V_{i+1}(x(t), t) \leq V_i(x(t), t), i = 1, 2, \ldots, p$;

*Then the equilibrium point is uniformly stable in the sense of Lyapunov.*

According to Theorem one and Theorem two, the stability conditions of a nonlinear switching system require that the value of the Lyapunov function is non-increased in the time interval of each continuous subsystem and non-increased in the whole switching process so that the energy function value is non-increased throughout the period. According to the analysis of Sect. 3.4, the mathematical model of integrated brake system for electromagnetic and friction in three braking modes is similar, and the stability analysis of the integrated brake system can be carried out by the standard Lyapunov function method. The switching control law and the Lyapunov function are:

$$\begin{cases} \Phi = -s \cdot sat\left(\lambda_{opt} - \lambda\right) \\ V = \left[\lambda_{opt}(t) - \lambda(t)\right]^2 \end{cases} \tag{6.2}$$

In the formula: $s$—Quantification factor; *Sat* $(x)$—saturation function theorem one condition one is satisfied, and the Lyapunov function is guided to get:

$$V = -2k\left(\lambda_{opt}(t) - \lambda(t)\right) \cdot sat\left(\lambda_{opt}(t) - \lambda(t)\right) \tag{6.3}$$

Since the saturation function also changes when $\lambda_{opt}(t) - \lambda(t)$ is changed, the constant is $\dot{V} < 0$ obtained, and condition two is satisfied. And theorem one in condition three proves as follows:

$$V_{i+1}(x(t)t) - V_i(x(t), t) = \left[\lambda_i(t) - \lambda_{i+1}(t)\right] \cdot \left[2\lambda_{opt}(t) - \lambda_{i+1}(t) - \lambda_i(t)\right] \tag{6.4}$$

In the formula: $[2\lambda opt(t) - \lambda i + 1(t) - \lambda i(t)] - 0 - [\lambda i(t) - \lambda i + 1(t)] < 0$. Thus, the formula (6.4) is less than 0 constantly established, and condition three in Theorem one is satisfied.

This shows that the switching control law and the standard Lyapunov function based on this paper make the integrated brake system stable during the switching between the three braking modes. The physical significance of this normal Lyapunov function is also apparent, requiring the integrated brake system for electromagnetic and friction to control the wheels during the brake mode switching vertical slip rate so that it is always lower than the optimum slip rate on the road surface, and the car brake is always in a stable area. Simultaneously, the stability of integrated brake system for electromagnetic and friction in braking mode switching is judged to be the difference between the best slip rate of the road surface and the vertical slip rate of the wheels is greater than or equal to zero. The car will not run off, side-slip, and other phenomena when the brake mode switches to ensure that the vehicle has better braking stability.

## *6.1.2 Controller Design of Braking Mode Switching Process*

### 6.1.2.1 Status Equation of Hybrid Brake-By-Wire System

From the point of view of system control, it is no longer meaningful to consider the dynamic characteristics of electronic, hydraulic brake when establishing the control model of integrated brake system for electromagnetic and friction, at this time mainly think the external factors of electronic, hydraulic brake and electromagnetic brake, that is, the influence of input control on the output braking moment. The expression of the braking moment of the electronic hydraulic brake output can be calculated in view of the pressure of the brake cylinder as follows:

$$T_h = K_{EHB} \cdot P_C \tag{6.5}$$

In the formula (6.5): $K_{EHB}$—Electronic, hydraulic brake gain factor, $K_{EHB} = 2\eta_c R_d A_C$;

At the same time, to further simplify the control model of the integrated brake system and ignore the effect of the nonlinear characteristics of the iron core and brake disc on the magnetic path analysis, the simplified (3.21) obtains the magnetic flux as follows:

$$\phi = \frac{NI}{\frac{\sqrt{2}\omega_d \Delta_h}{8\pi\rho_d} + \frac{l_g}{S_P \mu_0}} \tag{6.6}$$

According to the theory of vortex demagnetization effect, the electromagnetic braking moment is proportional to one-half of the speed of the brake disc. When the wheel speed is below a specific value, it can be assumed that there is a linear relationship between the electromagnetic braking moment and the wheel speed. Equation (3.22) is simplified to obtain:

$$T_e = K_{ECB} \cdot \omega \cdot I^2 \tag{6.7}$$

In the formula: $K_{ECB}$—electromagnetic brake gain factor.

The control model of the integrated brake system for electromagnetic and friction can be reduced to:

$$\begin{cases} \dot{\omega} = \frac{1}{J_W}\left(F_z \cdot K_\lambda \cdot \lambda \cdot R_{roll} - K_{EHB} P_C - K_{ECB}\omega I^2\right) \\ \dot{v} = -\frac{1}{M_V}(F_z \cdot K_\lambda \cdot \lambda) \end{cases} \tag{6.8}$$

there into: $K_\lambda = \frac{\varphi_h}{\lambda_{opt}}$.

#### 6.1.2.2 Reversibility Analysis of Hybrid Brake-By-Wire System

Next, the Interactor algorithm is used to prove the reversibility of integrated brake system for electromagnetic and friction. For the multi-input and multi-output integrated brake system for electromagnetic and friction described in the state Eq. (6.8), its state variable is $\boldsymbol{x} = [\omega, v]^T$, the input control quantity is $\boldsymbol{u} = [P_C, I]^T$, and the output variable is $\boldsymbol{y} = [F_{xb}, T_e]^T$. Calculate the derivative of the output variable $\boldsymbol{y}$ and the rank of the Jacobian matrix of the derivative equation of the output variable $\boldsymbol{y}$ to the input variable $\boldsymbol{u}$. When the derivative equation contains the input variable $\boldsymbol{u}$, the derivation is stopped.

Firstly, find the $\boldsymbol{r_1} = 1$ derivative of $\boldsymbol{y_1}$:

$$\dot{y}_1 = F_z \cdot R_{roll} \cdot K_\lambda \cdot \left(\frac{\dot{\omega}}{v} - \frac{\dot{v}}{v^2}\right) \tag{6.9}$$

Substitute Eq. (6.8) into Eq. (6.9) to obtain:

$$\dot{y}_1 = F_z R_{roll} \frac{K_\lambda}{J_W v}\left(F_z R_{roll} K_\lambda \lambda - K_{EHB} P_C - K_{ECB}\omega I^2\right) - F_z R_{roll} K_\lambda \frac{\dot{v}}{v^2} \tag{6.10}$$

Note $\boldsymbol{Y_1} = \boldsymbol{y_1}$, its rank to the Jacobian matrix of the input variable $\boldsymbol{u}$ is:

$$t_1 = rank\left[\frac{\partial Y_1}{\partial P_C}, \frac{\partial Y_1}{\partial I}\right] = rank\left[F_z R_{roll} K_\lambda \frac{K_{cb}}{J_W v}, F_z R_{roll} K_\lambda \frac{2K_{eb}\omega I}{J_W v}\right] = 1 \tag{6.11}$$

Secondly, find the $\boldsymbol{r_2}$ = 2nd derivative of $\boldsymbol{y_2}$:

$$y_2 = T_e = \frac{K_{ECB} I^2}{J_W}\left(F_z R_{roll} K_\lambda \lambda - K_{EHB} P_C - K_{ECB}\omega I^2\right) \tag{6.12}$$

Note $\boldsymbol{Y_2} = [Y_1, y_2]^{\mathrm{T}}$, and It's rank to the Jacobian matrix of the input variable $\boldsymbol{u}$ is:

$$t_2 = rank\begin{bmatrix} \frac{\partial Y_1}{\partial P_C}, \frac{\partial Y_1}{\partial I} \\ \frac{\partial Y_2}{\partial P_C}, \frac{\partial Y_2}{\partial I} \end{bmatrix} = \begin{bmatrix} F_z R_{roll} K_\lambda \frac{K_{EHB}}{J_W v}, F_z R_{roll} K_\lambda \frac{2K_{ECB}\omega I}{J_w v} \\ \frac{K_{ECB} I^2}{J_W} K_{EHB}, F_z R_{roll} K_\lambda \frac{\lambda K_{ECB}}{J_W} 2I - \frac{K_{ECB}^2 \omega}{J_w} 4I^3 \end{bmatrix} = 2 \tag{6.13}$$

Due to the existence of non-negative integers, $\boldsymbol{t_2}$ is equal to the number of system outputs. So the Interactor algorithm ends here.

$$\det\begin{bmatrix} \frac{\partial Y_1}{\partial P_C}, \frac{\partial Y_1}{\partial I} \\ \frac{\partial Y_2}{\partial P_C}, \frac{\partial Y_2}{\partial I} \end{bmatrix} = \begin{bmatrix} F_z R_{roll} K_\lambda \frac{K_{EHB}}{J_W v}, F_z R_{roll} K_\lambda \frac{2K_{ECB}\omega I}{J_W v} \\ \frac{K_{ECB} I^2}{J_W} K_{EHB}, F_z R_{roll} K_\lambda \frac{\lambda K_{ECB}}{J_W} 2I - \frac{K_{ECB}^2 \omega}{J_W} 4I^3 \end{bmatrix} \neq 0 \tag{6.14}$$

On the basis of the definition of vector relative order, the natural vector order and vector relative order of integrated brake system for electromagnetic and friction is the same, both of which are $\boldsymbol{a}$ = [1,3]T. Therefore, it can be concluded that the mathematical model of integrated brake system for electromagnetic and friction based on Eq. (6.8) is reversible. However, only when the system parameters are accurate, known, and kept constant can the pseudo linear system of integrated brake system for electromagnetic and friction be decoupled.

#### 6.1.2.3 Construction and Training of Neural Network Inverse System

The inverse system of integrated brake system for electromagnetic and friction is constructed by BP neural network and integrator. BP neural network is used to approximate the nonlinear mapping of the inverse system of integrated brake system for electromagnetic and friction, and integrator is used to characterize the dynamic characteristics of the inverse system. The input of the inverse system includes $\boldsymbol{F_{xb}}$ and $\boldsymbol{T_e}$ and two corresponding integrators, and the input layer of BP neural network

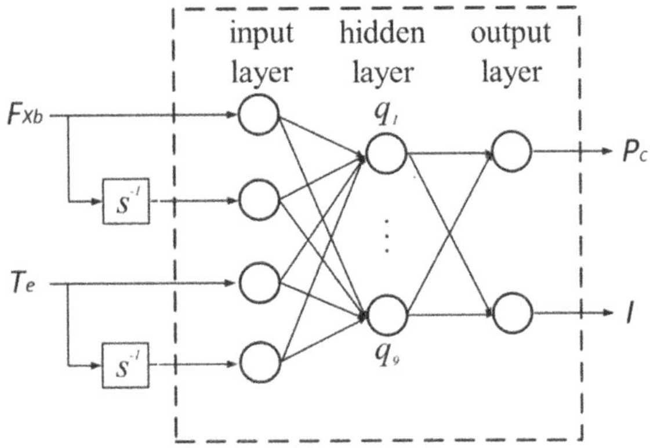


**Fig. 6.2** Structure diagram of the inverse system of the integrated brake system for electromagnetic and friction

contains four neurons; The output of the inverse system includes $\boldsymbol{P_C}$ and $\boldsymbol{I}$, then the output layer of BP neural network contains two neurons; According to the trial and error method, it can be determined that the hidden layer of BP neural network contains nine neurons, as Fig. 6.2 [3] shows.

The BP neural network model of the inverse system of integrated brake system for electromagnetic and friction is established by using the newff function in the Matlab/Simulink neural network toolbox. The input layer of BP neural network adopts purelin as the transfer function, the output layer adopts logsig as the transfer function, and the hidden layer adopts tansig as the transfer function. When testing BP neural network, firstly, tramnmx function is used to preprocess the data, and then postmnmx function is used to normalize the data. When training the BP neural network, the traindm function is selected as the training function, the training times are set to 300, the learning efficiency is 0.05, and the network target error is $10^{-3}$ After 45 times of training, the BP neural network reaches the required training accuracy.

#### 6.1.2.4 Design of Closed-Loop Controller of Hybrid Brake-By-Wire System

After decoupling the integrated brake system for electromagnetic and friction according to the neural network inverse system theory, the original integrated brake system for electromagnetic and friction with dual input and dual output is transformed into two non-interference single input and single output systems. As Fig. 6.3 shows, according to the feedback control structure of the integrated brake system for electromagnetic and friction, the electro-hydraulic brake performs the feedback control of the braking force of the friction brake through the pressure sensor installed at the brake wheel cylinder, and the electromagnetic braking feedback signal comes from the brake pedal travel sensor and the pressure sensor at the brake wheel cylinder, That is, the difference between the braking torque what the driver needs and the electro-hydraulic braking torque. The electro-hydraulic brake controller and electromagnetic brake controller can be designed by PID control, pole assignment or

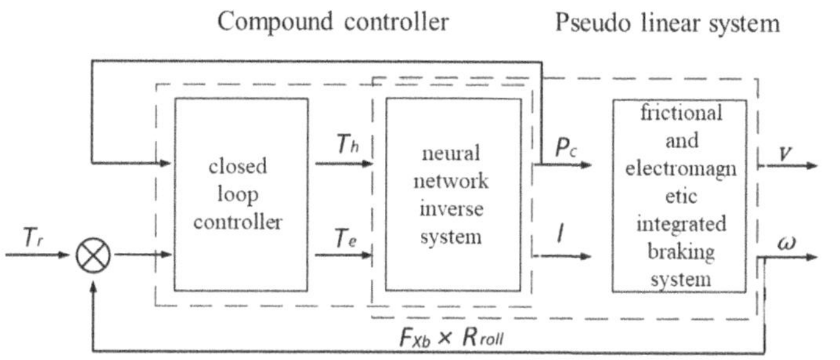


**Fig. 6.3** Block diagram of the decoupling controller of integrated brake system for electromagnetic and friction

quadratic index optimal control in the theory of univariate linear system, which will not be repeated here. Therefore, the design focus of the closed-loop controller is to propose the coordinated control strategy of the integrated brake system for electromagnetic and friction under ordinary braking conditions, which involves two kinds of practical problems to be solved: firstly, in the combined braking mode, the integrated brake system for electromagnetic and friction distributes the braking force of the front and rear axles on the basis of the ideal braking force distribution curve, The electro-hydraulic brake and electromagnetic brake on the front axle or rear axle need to be coordinated on the basis of the braking requirements of the driver; Secondly, in the process of braking mode switching, according to the switching control strategy of integrated brake system for electromagnetic and friction, it is indispensable to implement coordinated control between electro-hydraulic brake and electromagnetic brake to meet the changing braking needs of drivers.

These two kinds of practical problems are abstractly summarized. According to whether there is driver's braking intention intervention in the working process of integrated brake system for electromagnetic and friction, it is divided into two cases: coordinated control without driver's intention intervention and coordinated control with driver's intention intervention. When there is driver's intention intervention, an integrated brake system for electromagnetic and friction need to track and meet the driver's braking demand, Including the change of braking strength and change rate. At the same time, according to which braking mode is used in the coordinated control strategy of integrated brake system for electromagnetic and friction under ordinary braking conditions, it is divided into electro-hydraulic brake active type and electromagnetic brake active type.

### Active Coordinated Control Strategy of Electro-Hydraulic Brake

First of all, the electro-hydraulic braking active control cooperative control strategy of integrated brake system for electromagnetic and friction is proposed. Without the driver's intention to intervene, the electro-hydraulic braking torque is required to follow the change function of switching braking torque:

$$\begin{cases} T_h = f(T_s, \dot{T}_s) \\ T_e = T_s - T_h \end{cases} \tag{6.15}$$

where: $T_s$—switching braking torque.

Under the intervention of the braking intention of the driver, when the driver's braking intensity is lower than 0.1, stop using electro-hydraulic brake on the basis of the braking mode switching control strategy of integrated brake system for electromagnetic and friction proposed in Sect. 6.1.1, control electromagnetic brake to track the changing driver's required braking torque:

$$\begin{cases} T_e = f(T_r, \dot{T}_r) \\ T_h = 0 \end{cases} \tag{6.16}$$

where: $T_r$—braking torque needed for the driver.

When the driver's braking strength is higher than 0.1, control the electro-hydraulic brake to meet the changing driver's braking demand, while the electromagnetic braking torque remains constant:

$$\begin{cases} T_h = f(T_r, \dot{T}_r) \\ \dot{T}_e = 0 \end{cases} \tag{6.17}$$

Simultaneously, according to the stability criterion of integrated brake system for electromagnetic and friction in the process of braking mode switching obtained in Sect. 6.1.3, the combined braking torque and wheel longitudinal slip rate of electro-hydraulic brake are monitored in the process of braking mode switching. The electronic, hydraulic pressing power torque is controlled to make the wheel longitudinal slip rate consistently lower than the best slip rate on the road.

#### Active Coordinated Control Strategy of Electromagnetic Brake

Secondly, the active control coordinated control strategy of electromagnetic brake of integrated brake system for electromagnetic and friction is proposed. Without the intervention of the driver, the target braking torque of electro-hydraulic brake is controlled hierarchically, and the electromagnetic brake is used to follow and switch the change of braking torque:

$$\begin{cases} T_h = T_{s1}, T_{s2}, T_{s3}, \ldots \\ T_e = T_s - T_h \end{cases} \tag{6.18}$$

Under the intervention of the braking intention of the driver, electromagnetic brake shall be used to track the changing driver's required braking torque, and a threshold value of electromagnetic braking torque shall be set, which shall not exceed the threshold value; When the electromagnetic braking torque reaches the threshold value, control the electro-hydraulic brake to realize step-by-step pressure rise or pressure drop. Similarly, the combined braking torque and wheel longitudinal slip rate of electro-hydraulic and electromagnetic brake are monitored during the braking mode switching process. The electromagnetic braking torque is controlled to make the wheel longitudinal slip rate consistently lower than the best slip rate on the road.

## 6.1.3 Mode Switching Control of Composite Brake System for Motor and Friction

### 6.1.3.1 Braking Coordinated Control Method

In engineering practice, the braking energy recovery control strategy of fuel cell vehicles is based on rules. When the fuel cell vehicle is recovering the braking energy, to prevent the attenuation caused by the charging power of the power battery frequently exceeding its maximum charging power, the motor needs to recover the braking energy after the load reduction of PEMFC, which does not maximize the recovery of braking energy. Therefore, when designing the braking energy recovery control of fuel cell vehicle, this paper considers the dynamic load reduction characteristics of PEMFC to gradually increase the motor braking power, supplement the load reduction power of PEMFC, and ensure that the total charging power does not exceed the maximum charging power of power battery. Therefore, this paper takes the current control level of the existing motor. An electronically controlled pneumatic braking system as the premise, Aiming at recovering braking energy to the greatest extent and ensuring braking comfort during braking of fuel cell vehicle, based on the hierarchical control idea, We should design a coordinated control strategy for braking energy recovery of fuel cell vehicle considering the dynamic load reduction characteristics of PEMFC. A typical control method is established for the control strategy.

As shown in Fig. 6.4 [4, 5], We should determine the distribution of regenerative braking torque of the rear axle motor by combining the maximum charging power allowed by the present power battery and considering the motor power during PEMFC load reduction. The obtained motor demand braking torque command controls the motor through the motor controller. The distribution of braking torque required by the engine can be expressed as:

$$T_{mot\_req} = \min(T_{r\_req}, T_{r\_mot}) \tag{6.19}$$

where: $T_{mot\_req}$ is the required braking torque of the motor, $T_{r\_req}$ is the total braking torque demand value of the rear axle allocated by the braking energy recovery coordinated control strategy of the fuel cell vehicle considering the dynamic load reduction characteristics of PEMFC, and $T_{r\_mot}$ is the torque of the motor under the coordinated control strategy considering the dynamic load reduction characteristics of PEMFC,

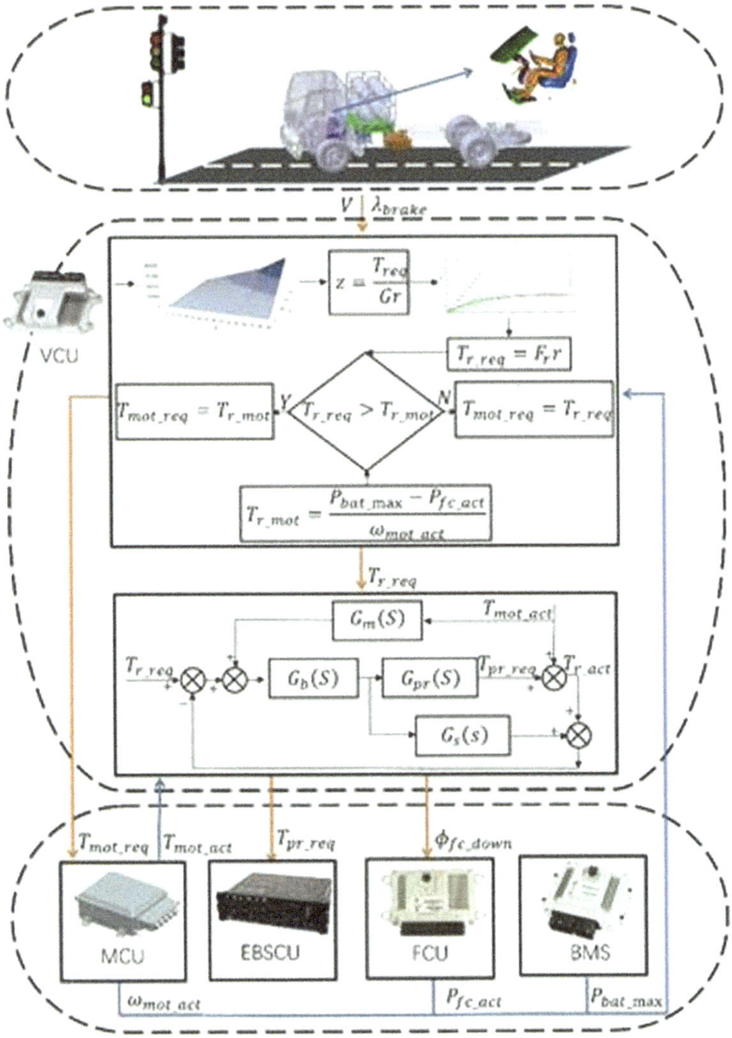


**Fig. 6.4** Coordinated control block diagram of braking energy recovery of fuel cell vehicle

which can be expressed as:

$$T_{r\_mot} = \frac{P_{bat\_\max} - P_{fc\_act}}{\omega_{mot\_act}} \tag{6.20}$$

where: $P_{bat\_\max}$ is the maximum charging power allowed by the power battery, and $P_{fc\_act}$ is the instantaneous power during PEMFC load reduction, $\omega_{\text{mot\_act}}$ is the angular speed of the motor.

In practical engineering applications, the hybrid brake system of fuel cell vehicles adopts thc braking energy recovery control method of motor regenerative braking torque to compensate the electronically controlled pneumatic braking torque, that is, the rule-based control method. As shown in Fig. 6.4, this paper proposes a coordinated control method of braking energy recovery based on the EBS air pressure system, which uses electronically controlled air pressure braking torque to compensate for motor regenerative braking torque, that is, a coordinated control method. This control method can randomly distribute the braking torque of front and rear axles, and the braking torque of the show and rear wheels can be distributed according to the I curve. We take the actual braking torque of the motor as the disturbance in the braking torque coordinated control algorithm. The motor braking torque demand value transmitted to the motor controller is the value allocated by means of the fuel cell vehicle braking energy recovery coordinated control strategy considering the dynamic load reduction characteristics of PEMFC, so it does not affect the braking energy recovery energy of the car. In the text, the transfer function GPR (s) of the air pressure system of fuel cell vehicle is a first-order lag system, which can be represented as:

$$G_{pr}(s) = \frac{C_{pr}}{\varphi_{pr}s + 1} \tag{6.21}$$

where: $\varphi_{pr}$ is the time constant of the barometric system, $C_{pr}$ is the barometric system constant and $s$ is a complex variable.

Setting the feedback channel aims to minimize external disturbance and model error on the control system. In the text, the feedback control based on PID control is used to eliminate the effect of model error on the control effect. As Fig. 6.4 shows, TDR is the braking torque calculated after adding feedforward control to the control system error

$$T_{dr} = T_{r\_req} - T_{r\_act} + T_{mot\_act}G_m(s) \tag{6.22}$$

where: $T_{r\_req}$ is the required total braking torque of the rear axle, $T_{r\_act}$ is the actual full brakimng torque of the rear axle, and $T_{mot\_act}$ is the actual braking torque of the motor, $G_m$(s) is the transmission of the feedforward channel of the control system.

The target of coordinated control is to differentiate between the required total braking torque and the actual total braking torque of the rear axle zero. According to this goal, the value of GM (s) transmitted by the feedforward channel is:

$$G_m(s) = -\frac{1}{G_b(s) \cdot G_{pr}(s)} \tag{6.23}$$

Thus, the braking torque required by the pneumatic system can be expressed as:

$$T_{pr\_req} = T_{dr} \cdot G_b(s) \cdot G_{pr}(s) \tag{6.24}$$

where: $G_b(s)$ is the transfer function of PID feedback channel, which can be expressed as:

$$G_b(s) = k_p + \frac{k_i}{s} + k_d s \tag{6.25}$$

where: $k_p$, $k_i$, $k_d$ are proportional gain, integral constant, and differential constant respectively.

To eliminate the lag link of the controlled object GPR (s) in the feedback signal and ensure the closed-loop stability of the control system, Smith predictive control transfer function GS (s) is added, which can be expressed as follows

$$G_s(s) = G_{pr}(1 - e^{-\tau s}) \tag{6.26}$$

where: $\tau$ is the lag time.

Figure 6.5 [6] is the physical connection diagram of the hardware in the loop (HIL) simulation system designed in this paper, composed of hardware and software. The hardware includes a programmable power supply, XPC host computer, target machine PXI system, integrated signal management unit, and electronic control original. The software uses VeriStand and MATLAB/Simulink to complete the design of the host computer control human–computer interaction interface and the modeling of the braking energy recovery system for fuel cell vehicles.

In order to examine the feasibility of the control strategy and control method of fuel cell vehicle braking energy recovery coordinated control considering PEMFC dynamic load reduction characteristics proposed in this paper, the control strategy

**Fig. 6.5** Physical connection diagram of HIL Simulation System

and method model of the braking energy recovery system for fuel cell vehicles are built-in Matlab/Simulink environment. The model is transformed into C code through an automatic code generation tool and burned into the vehicle controller. Connected to HIL, HIL simulation is conducted for the coordinated control strategy when considering the dynamic load shedding characteristics of PEMFC and the rule-based control strategy when not considering the dynamic load shedding parts of PEMFC under the urban, suburban, and intercontinental simulation conditions of WVU series cycle conditions, and the simulation data: fuel battery power, motor power, power battery power, and power battery SOC are compared and analyzed. Simultaneously, HIL simulation is carried out for the coordinated control method and rule-based control method of fuel vehicle braking energy recovery system under the urban, suburban and intercontinental simulation conditions of WVU series cycle conditions, and the simulation data: motor braking torque, pneumatic braking torque, and total braking torque are analyzed.

#### 6.1.3.2 Braking Coordinated Control Simulation

HIL Simulation Results and Analysis Under Urban Conditions

As Fig. 6.6 shows, the fuel cell vehicle starts braking at 677.5 S. as demonstrated in Fig. 6.6a, PEMFC starts load shedding. As Fig. 6.6b, c shows, when considering the dynamic load reduction characteristics of PEMFC, based on the coordinated control strategy, the motor power begins to increase during load reduction of PEMFC. The active battery is charged with the maximum charging power. Under the rule-based control strategy without considering the dynamic load shedding characteristics of PEMFC, the motor stops working when PEMFC is unloaded, and the power battery is charged with the power when PEMFC is unloaded. The HIL simulation results in Fig. 6.7b, c verify the correctness of the coordinated control strategy of the braking energy recovery system for fuel cell vehicle considering PEMFC dynamic load shedding characteristics proposed in this paper, as shown in Fig. 6.7d, At the end of the working condition, the SOC value of the fuel cell vehicle braking energy recovery coordinated control strategy considering the dynamic load reduction characteristics of PEMFC is 40.33, and the power cell SOC value under the rule-based control strategy without considering the dynamic load reduction characteristics of PEMFC is 39.83. Therefore, the recovery rate of vehicle braking energy under the coordinated control strategy is higher, and the fuel cell vehicle recovers more energy during braking.

Figure 6.7 shows the HIL simulation results of wheel braking torque under urban conditions. The fuel cell vehicle starts braking at 677.5 s. It can be seen that the braking torque of the rear wheel in Fig. 6.7a, b is coordinated and distributed by the regenerative braking torque of the motor and the pneumatic braking torque. The rule-based control method can compensate the pneumatic braking torque through the motor braking torque to meet the total braking torque demand. At the same time, it can better solve the hysteresis problem of air pressure establishment and improve the

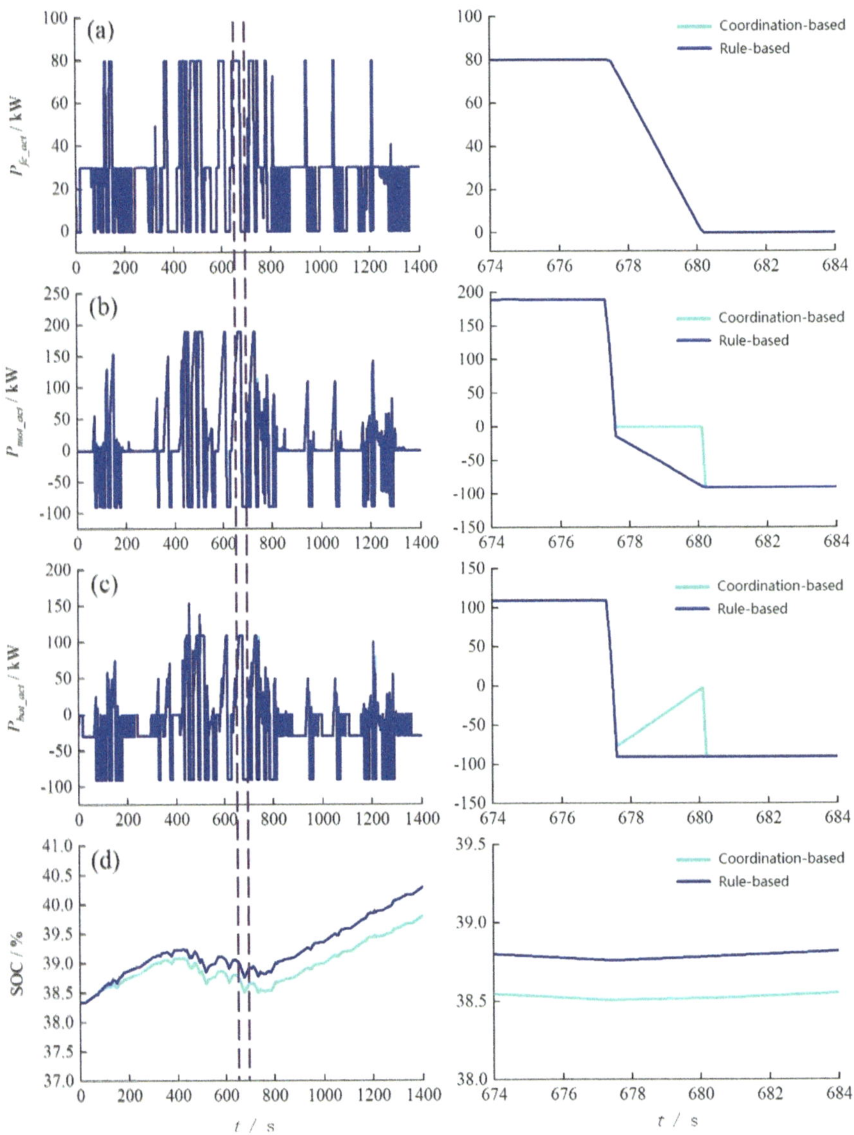


**Fig. 6.6** HIL simulation results under urban conditions

braking smoothness of the whole vehicle; The coordination-based control method used in this article can distribute the braking torque of the front-rear axles along the ideal braking curve. The pneumatic braking torque dynamically compensates the motor braking torque to meet the total braking torque demand. The motor braking torque can be released according to the dynamic load reduction characteristics of

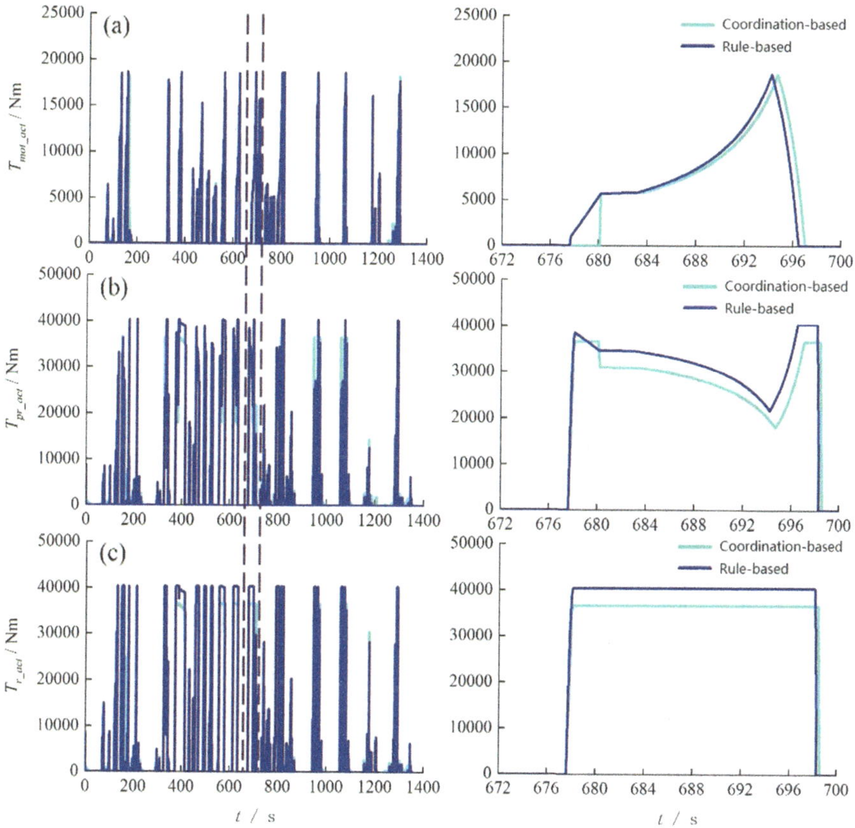


**Fig. 6.7** HIL simulation results of braking torque under urban working conditions

PEMFC to recover more energy, as shown in Fig. 6.8c. Compared with the rule-based control method, the maximum impact in fuel cell vehicles' braking energy recovery process is reduced by 3.2%. The braking comfort is improved to a certain extent.

#### HIL Smulation Results and Analysis Under Suburban Conditions

As Fig. 6.8 shows that the fuel cell vehicle starts braking at 508.6 S. as demonstrated in Fig. 6.8a, the PEMFC reduces the load. As Fig. 6.8b, c shows, when considering the dynamic load reduction characteristics of PEMFC, based on the coordinated control strategy, the motor power begins to increase during load reduction of PEMFC. The power battery is charged with the maximum charging power. Without considering the dynamic load reduction characteristics of PEMFC, under the rule-based control strategy, the motor stops working when PEMFC load reduction and the power battery is charged with the power when PEMFC load reduction. The HIL simulation results

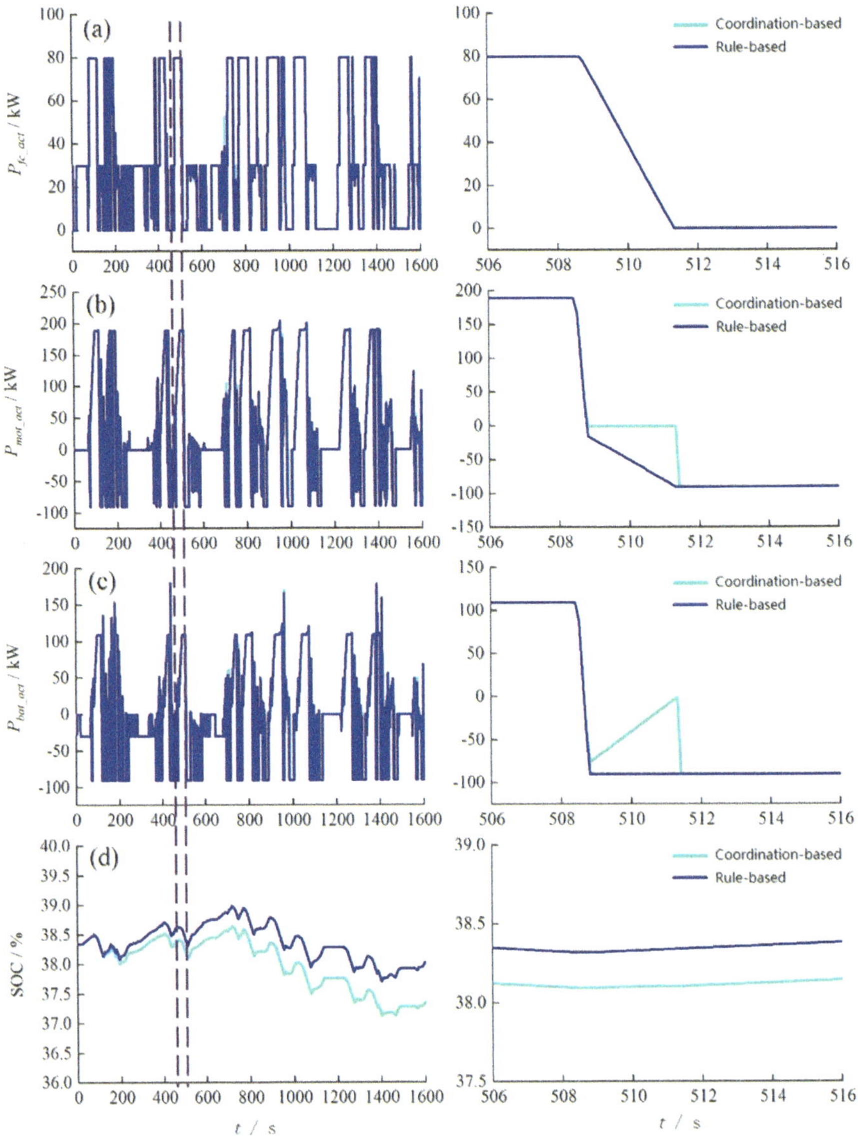


**Fig. 6.8** HIL simulation results under suburban conditions

in Fig. 6.8b, c verify the correctness of the coordinated control strategy of braking energy recovery system for fuel cell vehicle considering the dynamic load reduction characteristics of PEMFC proposed in this paper, as shown in Fig. 6.8d, At the end of the working condition, the SOC value of the fuel cell vehicle braking energy recovery coordinated control strategy considering the dynamic load reduction characteristics of PEMFC is 38.06. The power cell SOC value under the rule-based control strategy

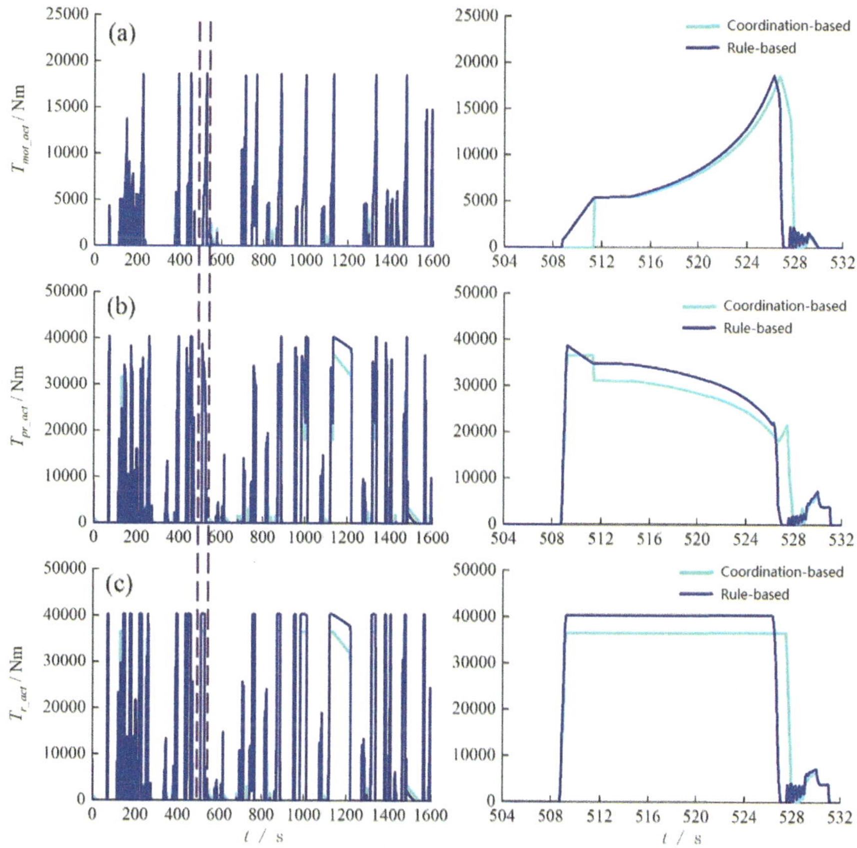


**Fig. 6.9** HIL simulation results of braking torque under suburban conditions

without considering the dynamic load reduction characteristics of PEMFC is 37.37. Therefore, the vehicle braking energy recovery rate under the coordinated control strategy is higher, and the fuel cell vehicle recovers more energy during braking.

Figure 6.9 shows the HIL simulation results of wheel braking torque under suburban conditions. The fuel cell vehicle starts braking at 508.6 s. It can be seen that the braking torque of the rear wheel in Fig. 6.9a, b is coordinated and distributed by the regenerative braking torque of the motor and the pneumatic braking torque. The rule-based control method can compensate the pneumatic braking torque through the motor braking torque to meet the total braking torque demand. At the same time, it can better solve the hysteresis problem of air pressure establishment and improve the braking smoothness of the whole vehicle; The coordination-based control method used in this paper can distribute the braking torque of the front, rear axles along the ideal braking curve. The pneumatic braking torque dynamically compensates the motor braking torque to meet the total braking torque demand. The motor braking torque can be released according to the dynamic load reduction characteristics of PEMFC to recover more energy, as Fig. 6.9c shows. Compared with the rule-based control method, the maximum impact in fuel cell vehicles' braking energy recovery process is reduced by 2.8%. The braking comfort is improved to a certain extent.

HIL Simulation Results and Analysis Under Intercontinental Conditions

As Fig. 6.10 shows, the fuel cell vehicle starts braking at 1362.2 s. As shown in Fig. 6.10a, the PEMFC reduces the load. As shown in Fig. 6.10b, c, when considering the dynamic load reduction characteristics of PEMFC, based on the coordinated control strategy, the motor power begins to increase during load reduction PEMFC. The power battery is charged with the maximum charging power. Without considering the dynamic load shedding characteristics of PEMFC, under the rule-based control strategy, the motor stops working when PEMFC is unloaded, and the power battery is charged with the power when PEMFC is unloaded. The HIL simulation results in Fig. 6.10b, c verify the correctness of the coordinated control strategy of braking energy recovery system for fuel cell vehicle considering PEMFC dynamic load shedding characteristics proposed in this paper, as shown in Fig. 6.10d, At the end of the working condition, the SOC value of the fuel cell vehicle braking energy recovery coordinated control strategy considering the dynamic load reduction characteristics of PEMFC is 37.75. The power cell SOC value under the rule-based control strategy considering the dynamic load reduction characteristics of PEMFC is 37.01. Therefore, the braking energy recovery rate of the whole vehicle based on the coordinated control strategy is higher, and the fuel cell vehicle recovers more energy during braking.

Figure 6.11 shows the HIL simulation results of wheel braking torque under suburban conditions. The fuel cell vehicle starts braking at 1362.2 s. It can be seen that the braking torque of the rear wheel in Figs. 6.11a, b is coordinated and distributed by the regenerative braking torque of the motor and the pneumatic braking torque. The rule-based control method can dynamically compensate the pneumatic braking torque through the motor braking torque to meet the total braking torque demand. At the same time, it can better solve the hysteresis problem of air pressure establishment and improve the braking smoothness of the whole vehicle; The coordinated control method can distribute the braking torque of the front and rear axles along the ideal braking curve. The pneumatic braking torque dynamically compensates the motor braking torque to meet the total braking torque demand. The motor braking torque can be released according to the dynamic load reduction characteristics of PEMFC to recover more energy, as Fig. 6.11c shows, compared with the rule-based control method, the maximum impact in the braking energy recovery process of fuel cell vehicle is reduced by 2.1%, and the braking comfort is improved to a certain extent.

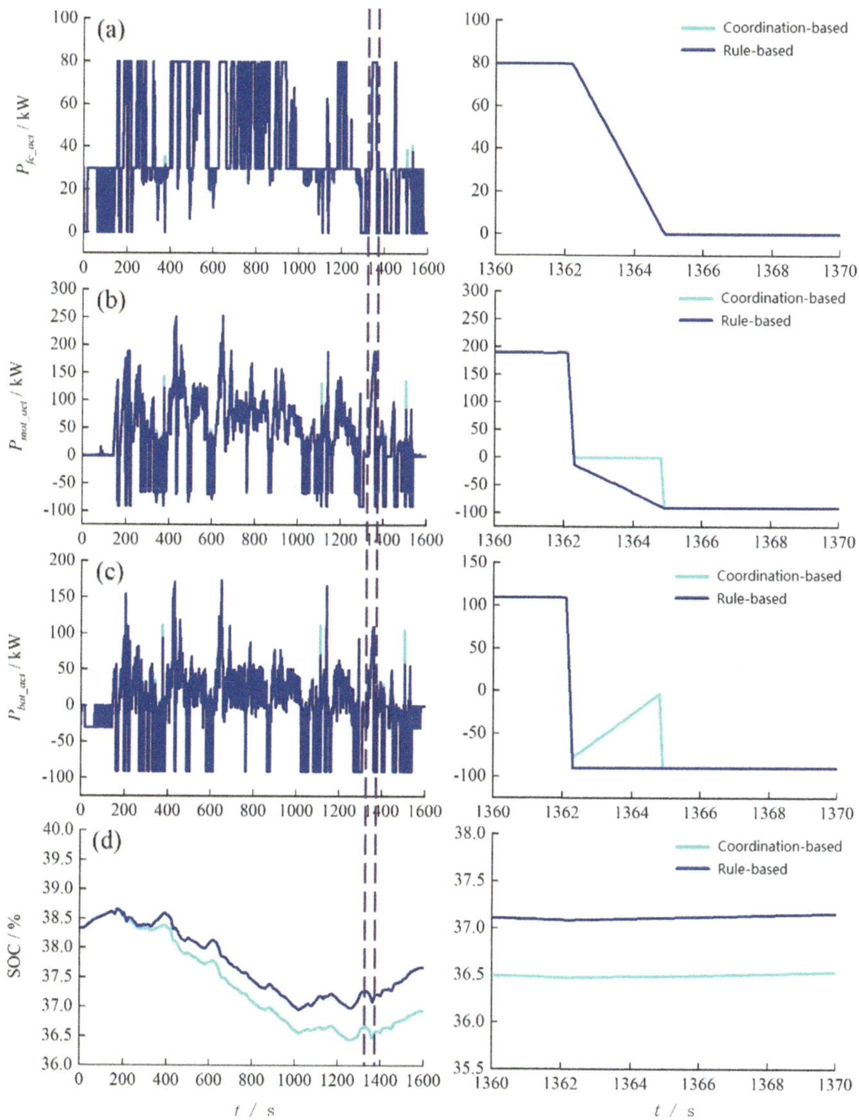


**Fig. 6.10** HIL simulation results under intercontinental conditions

### *6.1.4 Mode Switching Control of Integrated Brake System for Electromagnetic and Friction*

On the rapid control prototype platform of integrated brake system for electromagnetic and friction, firstly, based on the active coordinated control strategy of electro-hydraulic braking, the experimental research on braking mode switching of conflict and electromagnetic integrated brake system under three braking mode switching

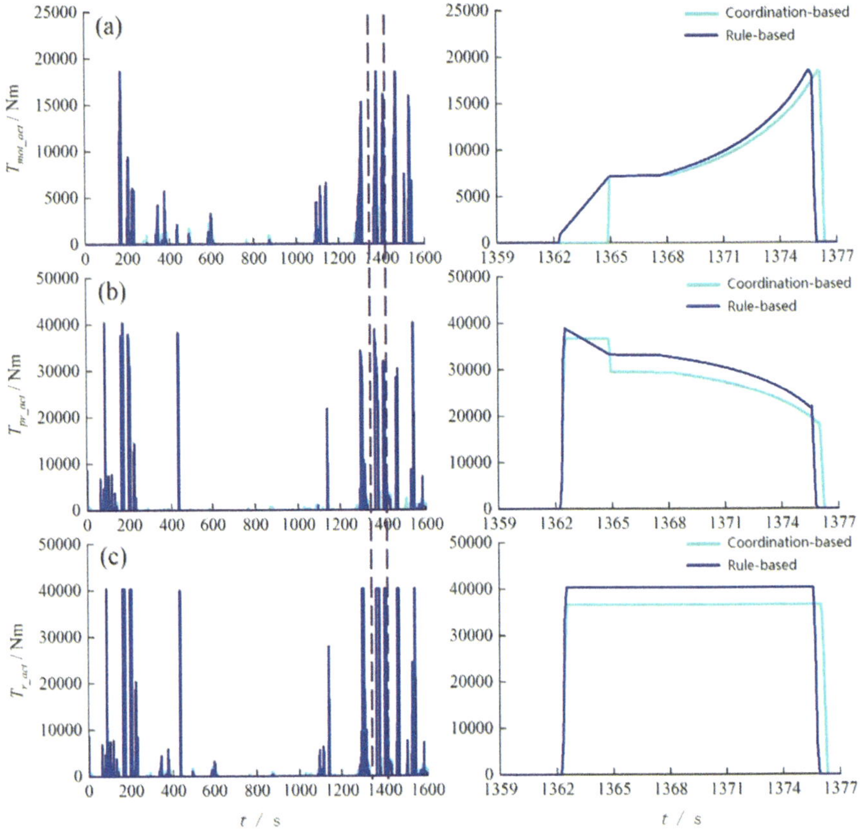


**Fig. 6.11** HIL simulation results of braking torque under intercontinental working conditions

conditions is carried out. The first working condition of braking mode switching is that the electromagnetic braking acts alone when the braking intensity is lower than 0.1. When the wheel speed decreases to 150 r/min, the electromagnetic braking shall be stopped to improve the service efficiency of the integrated brake system and switch the integrated brake system for electromagnetic and friction from the electromagnetic braking mode to the friction braking mode. As shown in Fig. 6.12, the wheel speed of the conflict and electromagnetic integrated brake system decreases to 150 r/min at 0.3 s, and the braking mode switching condition is triggered. The whole process from electromagnetic braking mode to friction braking model takes 0.2 s. The braking mode switching process is relatively stable. Still, the braking torque of the integrated brake system for electromagnetic and friction is slightly improved in the switching process, which is caused by the difficulty and poor control accuracy of electro-hydraulic braking under the demand of low braking intensity.

The second condition of braking mode switching is to use electromagnetic brake for independent braking in the initial state. As the driver steps on the brake pedal, the

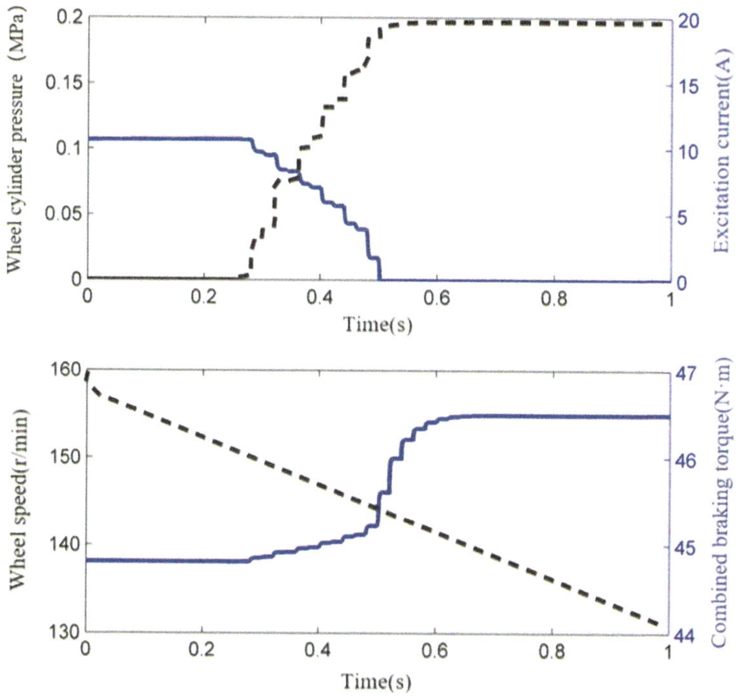


**Fig. 6.12** Experimental curves of the dynamic coordinated control for brake mode switch under the condition of driving cycle 1

required braking intensity increases, and the integrated brake system for electromagnetic and friction switch from electromagnetic braking mode to combined braking mode. As shown in Fig. 6.13, the expected braking strength of the driver starts to increase from 0.1 at 1 s and reaches 0.3 at 3 s. During the whole braking mode switching process, the electromagnetic braking torque can maintain the breaking strength of 0.1; At the initial stage of switching from electromagnetic braking mode to combined braking mode, the output electro-hydraulic braking torque has an unavoidable delay because the electro-hydraulic brake can not quickly respond to the braking demand.

The third condition of braking mode switching is to stop using electromagnetic brake to save energy consumption after the driver's required braking intensity remains unchanged for a while, and the integrated brake system for electromagnetic and friction is switched from combined braking mode to friction braking mode. As shown in Fig. 6.14, the electromagnetic braking stops at 0.4 s, the whole braking mode switching process lasts for 0.4 s, and the combined braking torque output is stable during this period. Under the switching conditions of three braking modes, the braking torque output by the integrated brake system for electromagnetic and

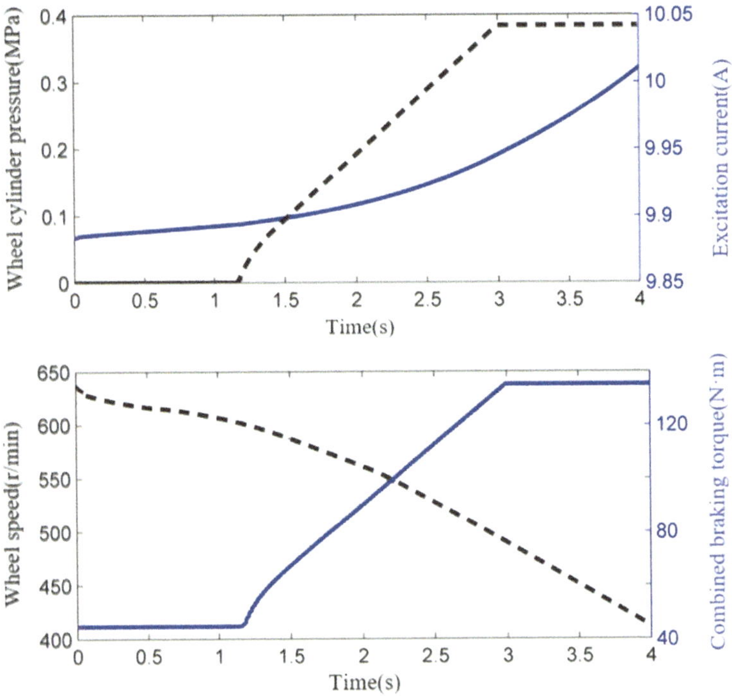


**Fig. 6.13** Experimental curves of the dynamic coordinated control for brake mode switch under the condition of driving cycle 2

friction well follows the braking demand of the driver, the wheel slip rate is lower than the optimal slip rate, and there is no instability.

Secondly, the experimental study of braking mode switching condition two is based on the active coordinated control strategy of electromagnetic brake, as shown in Fig. 6.15. At this time, the threshold value of electromagnetic braking torque is set to the braking torque corresponding to the breaking strength of 0.2. As shown in Figs. 6.13 and 6.15, based on the active coordinated control strategy of electro-hydraulic brake, the duration of the braking mode switching process is 2 s, and the natural frequency of the high-speed solenoid valve in electro-hydraulic brake is 50 Hz. To follow and meet the driver's braking demand, the high-speed solenoid valve needs to act 100 times in 2 s. In view of the active coordinated control strategy of electromagnetic brake, the high-speed solenoid valve works only four times in 2 s. Therefore, the operational cooperative control strategy based on electromagnetic brake can improve the control performance of integrated brake system for electromagnetic and friction, reduce the action frequency of high-speed solenoid valves, and improve the reliability of electro-hydraulic brake.

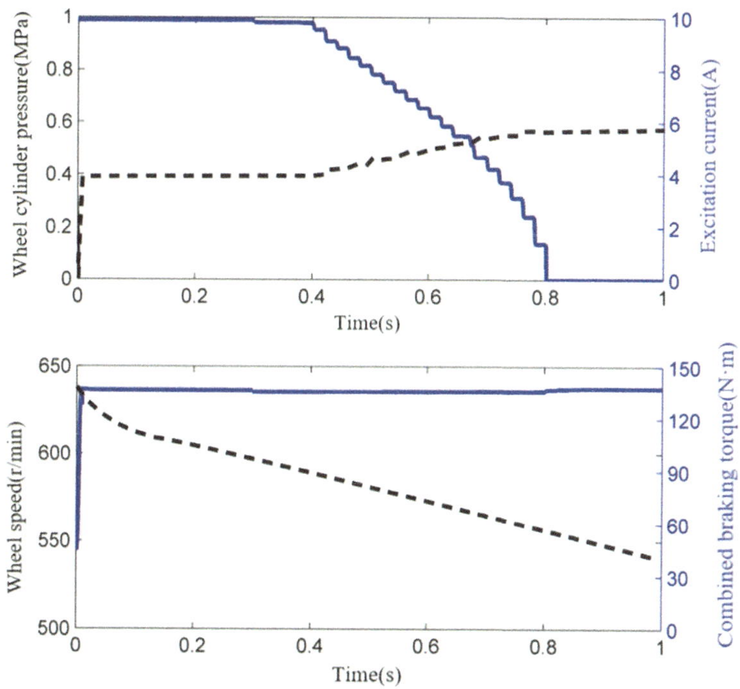


**Fig. 6.14** Experimental curves of the dynamic coordinated control for brake mode switch under the condition of driving cycle 3

## 6.2 Mode Switching Control of Hybrid Brake-By-Wire System Under Emergency Braking

### *6.2.1 Antilock Braking Modeling*

In the process of electro-hydraulic brake pressurization, the pressure difference from the accumulator to the brake wheel cylinder is equal to the sum of the pressure differences between the accumulator and the oil inlet valve and between the oil inlet valve and the brake wheel cylinder. Firstly, according to Eq. (3.12), the pressure difference between the brake pipeline and the brake wheel cylinder is:

$$P_{val} - P_C = R_H Q_C + L_H \frac{dQ_C}{dt} \tag{6.27}$$

Then, Eq. (3.10) is simplified, and the oil inlet and outlet of the high-speed solenoid valve are equivalent to an orifice, so the pressure-flow equation of the high-speed solenoid valve is:

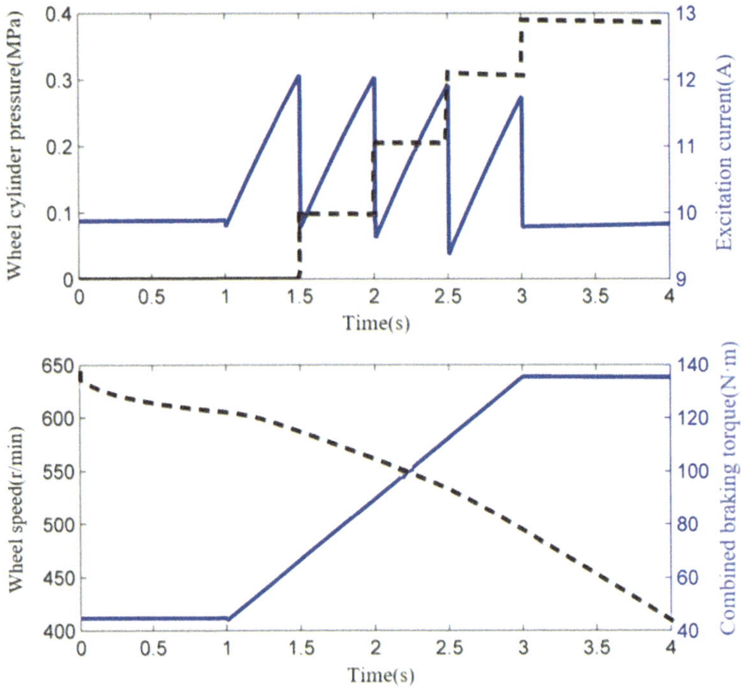


**Fig. 6.15** Experimental curves of the dynamic coordinated control for brake mode switch based on electromagnetic brake active control under condition of driving cycle 2

$$P_A - P_{val} = \frac{\rho_f}{2}\left(\frac{Q_A}{C_d A_{vi}}\right)^2 + \frac{\rho_f}{2}\left(\frac{Q_C}{C_d A_{vo}}\right)^2 \tag{6.28}$$

For simultaneous (6.27) and (6.28), the pressure difference between the accumulator and the brake wheel cylinder can be expressed as:

$$P_A - P_C = \frac{\rho_f}{2}\left(\frac{Q_A}{C_d A_{vi}}\right)^2 + \frac{\rho_f}{2}\left(\frac{Q_C}{C_d A_{vo}}\right)^2 + R_H Q_C + L_H \frac{dQ_C}{dt} \tag{6.29}$$

When establishing the hydraulic anti-lock braking model, we generally ignore the effects of hydraulic resistance and hydraulic sensing of the braking pipeline; At the same time, compared with the pressure change in the brake wheel cylinder, the gas pressure of the accumulator changes slowly, which can also be ignored. After simplifying Eq. (6.29), substitute Eq. (3.15) into Eq. (6.29) to obtain:

$$P_A - P_C = \frac{\rho_f}{2}\left(\frac{1}{C_d A_{vo}} \cdot \frac{V_C}{K_C} \cdot \frac{dP_C}{dt}\right)^2 \tag{6.30}$$

Equation (6.30) is a simplified model of electro-hydraulic brake and anti-lock braking. Considering the demand for model accuracy in further research work, the model needs to be treated equivalently. Firstly, the accumulator hydraulic capacity, brake pipeline hydraulic capacity, and brake wheel cylinder hydraulic capacity are superimposed to form the equivalent hydraulic degree of electro-hydraulic brake. Secondly, the accumulator hydraulic resistance, brake pipeline hydraulic resistance, and high-speed solenoid valve hydraulic resistance are superimposed to form the equivalent hydraulic resistance of electro-hydraulic brake. By adopting the above model correction method, Eq. (6.30) can be expressed as:

$$P_A - P_C = \frac{\rho_f}{2}\left(R_E C_E \cdot \frac{dP_C}{dt}\right)^2 \tag{6.31}$$

where: $R_E$—equivalent hydraulic resistance of electro-hydraulic brake, N·s/m$^5$; $C_E$—Equivalent fluid volume of electro-hydraulic brake, m$^3$/N.

Deform Eq. (6.31) to obtain the pressure equation of brake wheel cylinder when the electro-hydraulic brake is pressurized as follows:

$$\frac{dP_C}{dt} = \frac{1}{R_E C_E}\sqrt{\frac{2(P_A - P_C)}{\rho_f}} \tag{6.32}$$

Similarly, we can get the pressure equation of brake wheel cylinder during electro-hydraulic brake decompression as follows:

$$\frac{dP_C}{dt} = -\frac{1}{R_E C_E}\sqrt{\frac{2P_C}{\rho_f}} \tag{6.33}$$

When studying the anti-lock braking control method, we generally assume that the vehicle speed remains unchanged and only consider the change of wheel speed. Therefore, we can simplify the vehicle braking model represented by Eq. (3.23) and ignore the whole vehicle equation in the model. Then Eq. (3.23) can be expressed as:

$$\dot{\lambda} = -\frac{R_{roll}}{J_W}\left[F_z \cdot \mu(\lambda) \cdot R_{roll} - T_e - T_h\right] \tag{6.34}$$

Substituting Eqs. (6.5) and (6.7) into Eq. (6.34), the state equation of integrated brake system for electromagnetic and friction in the stable region is obtained as follows:

$$\begin{cases} \dot{P}_C = u_1 \cdot \frac{1}{R_E C_E}\sqrt{\frac{2(P_A - P_C)}{\rho_f}} - u_2 \cdot \frac{1}{R_E C_E}\sqrt{\frac{2P_C}{\rho_f}} \\ \dot{\lambda} = \frac{R_{roll}}{J_W}\left[(F_z K_\lambda \lambda R_{roll} + K_{ECB}\frac{v}{R_{roll}} I^2)\lambda - K_{ECB}\frac{v}{R_{roll}} I^2 - K_{EHB} P_C\right] \end{cases} \tag{6.35}$$

where:, $u_1$, $u_2$—duty cycle control signal of inlet valve and outlet valve.

Similarly, the state equation of the integrated brake system for electromagnetic and friction in the unstable region is obtained as follows:

$$\begin{cases} \dot{P}_C = u_1 \cdot \frac{1}{R_E C_E}\sqrt{\frac{2(P_A - P_C)}{\rho_f}} - u_2 \cdot \frac{1}{R_E C_E}\sqrt{\frac{2P_C}{\rho_f}} \\ \dot{\lambda} = \frac{R_{roll}}{J_W}\Big[(K_{ECB}\frac{v}{R_{roll}}I^2 - F_z K_m R_{roll})\lambda + F_z \mu_m R_{roll} \\ \qquad -K_{ECB}\frac{v}{R_{roll}}I^2 - K_{EHB}P_C\Big] \end{cases} \tag{6.36}$$

## 6.2.2 Antilock Braking Coordinated Controller

According to the layered coordinated control (LCC) theory, we guide the anti-lock braking coordination controller design of integrated brake system for electromagnetic and friction. As Fig. 6.16 shows, the anti-lock braking coordination controller includes an upper layer coordination controller and a lower layer sub-controller (electromagnetic brake controller and electro-hydraulic brake controller). The signal processing module obtains the vehicle driving state information according to the vehicle information sent by the sensor (estimation of vehicle speed, estimation of slip ratio, etc.); The signal processing module sends the vehicle driving state information to the upper coordination controller in real-time, including the speed of four

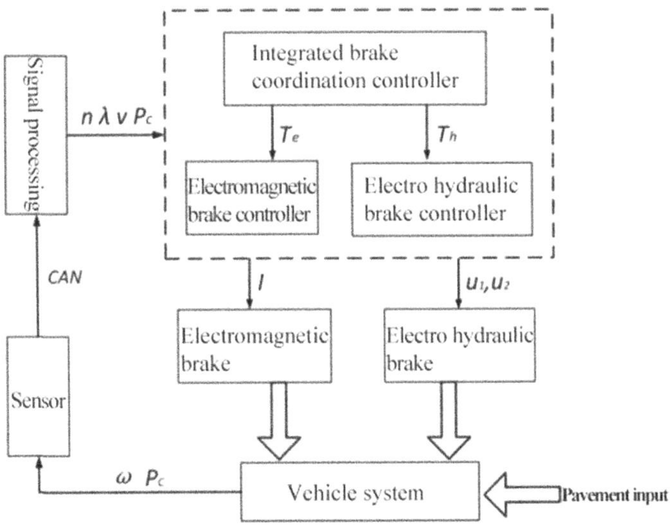


**Fig. 6.16** Block diagram of the hierarchical coordinated control of anti-lock braking for the integrated brake system for electromagnetic and friction

wheels, brake wheel cylinder pressure and slip rate, vehicle speed, etc.; According to the vehicle driving state information and the anti-lock braking coordination control strategy of integrated brake system for electromagnetic and friction, the upper coordination controller calculates the electromagnetic brake target braking torque and electro-hydraulic brake target braking torque in real-time and sends them to the electromagnetic brake controller and electro-hydraulic brake controller respectively. The electromagnetic brake controller controls the excitation current in electromagnetic brake on the basis of the target braking torque of electromagnetic braking; Similarly, the electro-hydraulic brake controller controls the duty cycle of the inlet valve and outlet valve in the electro-hydraulic brake according to the target braking torque of the electro-hydraulic braking.

#### 6.2.2.1 Upper Layer Coordination Controller

The demand of anti-lock braking control is separated and divided into two categories: one is to track and meet the demand of braking strength, including the estimation of pavement peak adhesion coefficient and the control aiming at braking strength; The second is the demand to realize the optimal slip rate control, including the assessment of the optimal slip rate and the power aiming at the optimal slip rate. Considering that compared with electro-hydraulic brake, electromagnetic brake has the advantages of rapid braking response and simple control, this paper puts forward a working idea different from the anti-lock braking control of traditional hydraulic brake system: using electromagnetic brake to replace electro-hydraulic brake to realize the demand of wheel optimal slip rate control, and electro-hydraulic brake is mainly used to track and meet the demand of braking strength. The anti-lock braking coordination control strategy of integrated brake system for electromagnetic and friction is as follows:

(1) Firstly, we should note that electromagnetic brake has apparent advantages in braking response speed. When current is applied to the excitation coil of the electromagnetic brake, electromagnetic brake starts to work, and the response time of electromagnetic brake is about 30–50 ms; After opening the hydraulic inlet valve, the high-pressure brake fluid in the accumulator enters the brake wheel cylinder through the hydraulic inlet valve and brake pipeline, and the friction brake generates braking torque. Compared with traditional hydraulic brake, the response time of electronic and hydraulic brake is shortened to about 0.2S. Secondly, we should point out that electro-hydraulic brake has apparent advantages in providing braking strength. The peak electromagnetic braking torque output by a single electromagnetic brake is about 200 N. M, while the braking pressure of friction brake in electro-hydraulic brake can reach the maximum working pressure of accumulator (about 16 MPa). Therefore, in the first stage of anti-lock braking, from the perspective of improving the anti-lock braking response time and reducing the vehicle emergency braking distance, the integrated brake system for electromagnetic and friction should make the electro-hydraulic brake and electromagnetic brake work together,

and the electromagnetic brake is required to output the peak electromagnetic braking torque at this time. If the electro-hydraulic brake eliminates the brake clearance of the friction brake and fully responds, the first stage of anti-lock braking ends.

(2) In the second stage of anti-lock braking, considering that electromagnetic brake can not fully meet the braking strength requirements under general emergency braking conditions, electronic, hydraulic brake is used uniformly to estimate the peak adhesion coefficient of the road; At the same time, the four wheels will be subject to ample electro-hydraulic braking torque after the electro-hydraulic complete response, which will lead to the rapid reduction of wheel speed and significant fluctuation of electromagnetic braking torque. Therefore, in the second stage of anti-lock braking, we should temporarily stop electromagnetic brake.

The expression of pavement adhesion coefficient obtained by deforming Eq. (3.23) is as follows:

$$\varphi(\lambda) = \frac{J_W \dot{\omega}(t) + T_h}{F_z R_{roll}} \tag{6.37}$$

Assuming that the sampling time of the grating sensor for the wheel speed is, the estimation expression of the wheel speed is:

$$\dot{\omega}(t) = \frac{\omega(t + \Delta t) - \omega(t)}{\Delta t} \tag{6.38}$$

The estimation expression of pavement peak adhesion coefficient obtained by combining formula (6.37) and formula (6.38) is:

$$\varphi(t) = \frac{(\omega(t + \Delta t) - \omega(t))J_W}{F_z R_{roll} \Delta t} + \frac{K_{EHB} P_C(t)}{F_z R_{roll}} \tag{6.39}$$

In Eq. (6.39), the pressure of the brake wheel cylinder can be obtained in real-time through the pressure sensor. Therefore, we can use electro-hydraulic brake to estimate the peak adhesion coefficient of pavement successfully. At this time, the second stage of anti-lock braking is completed.

(3) In the third stage of anti-lock braking, we hope that electro-hydraulic braking can provide the braking strength required by vehicle anti-lock braking control and keep the target braking torque constant, and then use electromagnetic brake alone to estimate the optimal slip rate of the road surface and control the actual slip rate of the wheel with the optimal slip rate as the target. The target braking torque of electro-hydraulic brake shall be equal to the braking torque corresponding to the peak adhesion coefficient of the road minus the expected output electromagnetic braking torque. The expression of the target braking torque of electro-hydraulic brake is:

$$T_h = \varphi_h F_z R_{roll} - 0.5 T_{e\max} \tag{6.40}$$

where: $T_{emax}$ is the peak electromagnetic braking torque, T.

As Fig. 3.8 shows, in the vehicle braking stability area, the slope of the road adhesion coefficient and slip ratio curve is positive; In terms of vehicle braking instability, the slope of the curve between road adhesion coefficient and slip ratio is negative; At the optimum slip ratio, the slope of the angle between pavement adhesion coefficient and slip ratio is zero. Therefore, according to the characteristics of the curve shape of pavement adhesion coefficient and slip ratio, the estimation expression of the optimal slip ratio can be obtained as follows:

$$\lambda(t + \Delta t) = \lambda(t) + \Phi \cdot \mathrm{sgn}\left(\frac{d\varphi}{d\lambda}\right) \tag{6.41}$$

where: $\Phi$—step constant.

In Eq. (6.41), the estimation expression of pavement adhesion coefficient and slip rate curve slope is:

$$\frac{d\varphi}{d\lambda} = \frac{\varphi(t + \Delta t) - \varphi(t)}{\lambda(t + \Delta t) - \lambda(t)} \tag{6.42}$$

Combined Eqs. (6.41) and (6.42) can estimate the optimal slip rate of the road surface. Further, electromagnetic brake can be used to control the actual slip rate of the wheel with the objective of the optimal slip rate.

(4) During vehicle anti-lock braking control process, if the jump of road adhesion conditions occurs, the single use of electromagnetic brake can not track and control the optimal slip rate of the road, and the actual slip rate of the wheel will also jump; Therefore, it is necessary to return from the third stage of anti-lock braking to the second stage of anti-lock braking and reuse electro-hydraulic brake to estimate the peak adhesion coefficient of the existing road surface.

#### 6.2.2.2 Lower Layer Sub-Controller

According to the analysis in Chap. 3, the integrated brake system for electromagnetic and friction is a coupled nonlinear system. At the same time, on the basis of the anti-lock braking coordination control strategy of the conflict and electromagnetic integrated brake system formulated in Sect. 7.2.1, the target braking torque of electro-hydraulic brake in the third stage anti-lock braking remains constant. Therefore, in this chapter, the coupling characteristics between electro-hydraulic brake and electromagnetic brake are ignored. Simultaneously, to consider the nonlinear aspects of electro-hydraulic brake and electromagnetic brake, the sliding mode variable structure theory (SVS) is applied to design the electromagnetic brake controller and electro-hydraulic brake controller, respectively.

Sliding Mode Switching Surface

The sliding mode variable structure theory is used for designing the electromagnetic brake controller and electro-hydraulic brake controller. Its main goal is to find the control law of electro-hydraulic brake and electromagnetic brake in the anti-lock braking process to make sure the stability of the control system and the dynamic characteristics that meet the performance index requirements. Considering the physical meaning of the anti-lock braking control system, when the actual wheel slip rate tends to the optimal slip rate, that is, the tracking error approaches zero. Once the system state variable enters the sliding mode switching surface, it can stably approach the state origin. Therefore, the sliding mode switching surface of the electromagnetic brake controller and the electro-hydraulic brake controller is defined as:

$$S = \lambda_{opt} - \lambda \tag{6.43}$$

Equivalent Control Quantity

During Anti-lock braking control on the switching surface, the braking torque output by the integrated brake system for electromagnetic and friction is discontinuous, which can be equivalent replaced by the average value in a sense, so that the system moves along the switching surface of = 0, which can ensure the generation of sliding mode:

$$\dot{S} = \dot{\lambda}_{opt} - \dot{\lambda} = 0 \tag{6.44}$$

The equivalent control quantity of electro-hydraulic brake can be expressed as:

$$P_C = \frac{1}{K_{EHB}}\left[\dot{\lambda} \cdot \frac{J_W v}{R_{roll}} + (F_z \cdot \frac{\mu_h}{\lambda_{opt}} \cdot R_{roll} + K_{ECB} \cdot \frac{v}{R_{roll}} \cdot I^2) \cdot \lambda\right] \tag{6.45}$$

The equivalent control quantity of electromagnetic brake can be expressed as:

$$I = \sqrt{R_{roll} \frac{F_z K_\lambda \lambda + \dot{\lambda} \cdot \frac{J_W v}{R_{roll}} - K_{EHB} P_C}{(1 - \lambda) \cdot v \cdot K_{ECB}}} \tag{6.46}$$

Accessibility Analysis of Switching Surface

When the system state variable is not on the sliding mode switching surface, we need to add a control term to make the phase trajectory move in sliding mode switching. Firstly, the expression of electro-hydraulic brake and electromagnetic brake control quantity is defined as:

$$u = u_{BL} - \kappa \operatorname{sgn}(S) \tag{6.47}$$

Under the condition of generalized sliding mode, the switching function of the sliding mode controller designed according to the constant speed approach rate () should meet the following requirements;

$$S\dot{S} \le -\eta|S|, \quad \eta > 0 \tag{6.48}$$

Combined formula (6.47) and formula (6.44) obtain:

$$S\dot{S} = -S\left(\frac{R_{roll}}{J_W}\left[F_Z \cdot \mu(\lambda) \cdot R_{roll} - u_{BL} - \kappa \operatorname{sgn}(S)\right]\right) \tag{6.49}$$

Then there

$$S\dot{S} = -S\left(\frac{R_{roll}}{J_W} \cdot \kappa \operatorname{sgn}(S)\right) \tag{6.50}$$

To ensure the accessibility of the switching surface, make:

$$\kappa = \frac{J_W}{R_{roll}}\eta \tag{6.51}$$

Substituting Eq. (6.51) into Eq. (6.50), we can get:

$$S\dot{S} \le -\eta|S|, \quad \eta > 0 \tag{6.52}$$

The above formula shows that the system satisfies the generalized sliding mode condition conditions and the existence and accessibility of the sliding mode switching surface.

### Method of Weakening System Chattering

The discontinuous switching function is used to express electro-hydraulic brake and electromagnetic brake control quantity represented by Eq. (6.47), which will cause high-frequency chattering of the system near the sliding mode switching surface. This high-frequency chattering will damage the executive parts of the control system, produce self-excited motion, and cause system instability. Based on the demand of weakening the high-frequency chattering of the system near the sliding mode switching surface, the saturation function is used instead of the symbolic role and substituted into the Eq. (6.47). The expression of electro-hydraulic brake and electromagnetic brake control quantity is as follows:

$$u = u_{BL} - \kappa sat\left(\frac{S}{\Lambda}\right) \tag{6.53}$$

Of which: $sat(\frac{S}{\Lambda}) = \begin{cases} \frac{S}{\Lambda}, |S| \le \Lambda \\ sign(S), |S| \le \Lambda \end{cases}$ Where: $\Lambda$—controller parameter, indicating the degree of leaving the ideal switching surface.

Then in the second stage of anti-lock braking, the equivalent control quantity of electro-hydraulic brake is:

$$\begin{aligned} P_C = \frac{1}{K_{EHB}}\Bigg[\eta \cdot \frac{J_W v}{R_{roll}} \cdot sat\left(\frac{\lambda_{opt} - \lambda}{\Lambda}\right) \\ +(F_z \cdot \frac{\mu_h}{\lambda_{opt}} \cdot R_{roll} + K_{ECB} \cdot \frac{v}{R_{roll}} \cdot I^2) \cdot \lambda\Bigg] \end{aligned} \tag{6.54}$$

In the third stage of anti-lock braking, the equivalent control quantity of electromagnetic brake is:

$$I = \sqrt{R_{roll}\frac{\left[F_z K_\lambda \lambda + \eta \cdot \frac{J_W v}{R_{roll}} \cdot sat\left(\frac{\lambda_{opt} - \lambda}{\Lambda}\right) - K_{EHB} P_C\right]}{(1 - \lambda) \cdot v \cdot K_{ECB}}} \tag{6.55}$$

### *6.2.3 Antilock Braking Hardware-In-The-Loop*

Hardware in the loop (HIL) simulation is an advanced testing method, which can test a large number of newly developed automotive electronic control units in a virtual environment without testing in a real vehicle. The advantages of this test method are: simulating various working conditions of the controlled object, simulating complex fault modes, quickly reproducing fault modes, realizing integrated test of multiple electronic control units, easy maintenance and expanding test capability, etc. As shown in Fig. 6.17 [7], the hardware in the loop simulation platform of integrated brake system for electromagnetic and friction is a set of hardware in the loop simulation system based on Ni company's software and hardware products and peripheral software and hardware such as conditioning, fault injection and load simulation, including real-time simulator hardware, test engineering management and simulation model.

#### 6.2.3.1 Real-Time Simulator Hardware

Real-time simulator hardware is the platform and carrier of software operation, which usually includes real-time hardware and peripheral hardware. The main functions

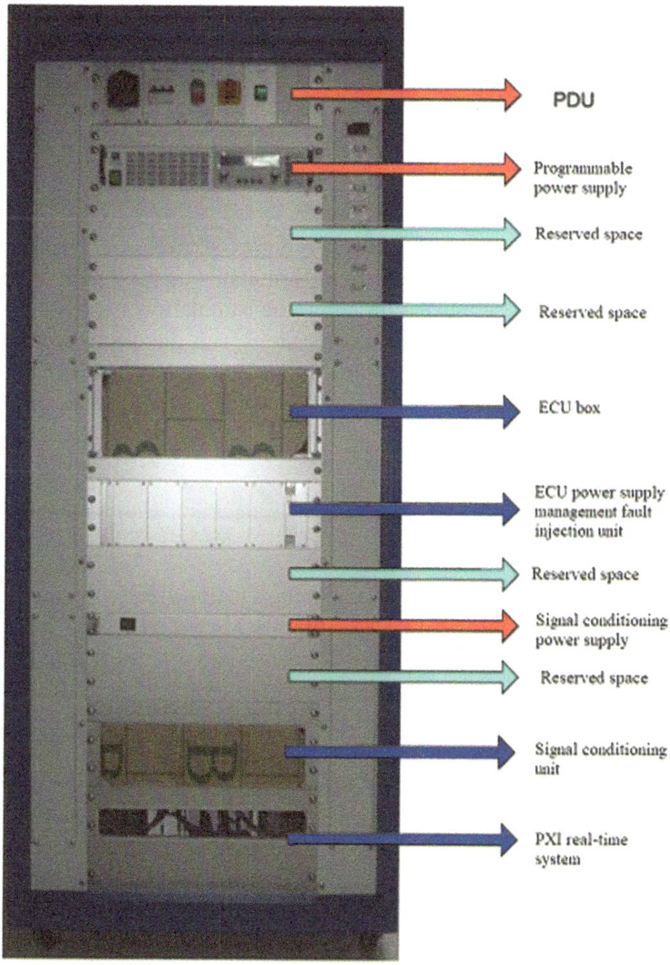


**Fig. 6.17** Hardware in the loop simulation platform of integrated brake system for electromagnetic and friction

of real-time hardware include the following three points: providing the hardware environment for the operation of the real-time model, that is, real-time processor and real-time operating system; The results calculated by the simulation model are converted into real physical signals to realize real-time I/O output; Collect the physical quantities of the real state of the reaction actuator, such as voltage, current, displacement, speed, etc., and convert them into parameters that can be identified by the simulation model. Peripheral hardware provides driving (such as voltage amplification, current amplification, etc.), protection (such as short-circuit protection, overvoltage, overcurrent protection, etc.)and signal conversion for real-time I/O signals(such as voltage to current conversion, current to voltage conversion, level conversion, etc.) a fully configured real-time simulator hardware mainly includes PXI real-time system, signal conditioning unit, fault injection unit, onboard power simulator, power distribution module, etc.

PXI Real-Time System

The PXI real-time system accepts various control signals sent by the electronic control unit through the hardware interface. Simultaneously, outputs various sensor signals to the electronic control unit after the simulation model calculation is completed, thereby forming a closed-loop system with the electronic control unit. The real-time boards in the PXI real-time system include Ni chassis (PXI-1042), Ni real-time processor (PXI-8108rt), FPGA boards(PXI-7831r), CAN board (PXI-8512 / 2).

Signal Conditioning Unit

The main functions of the signal conditioning unit include the following four points: allocate the hardware resources of the real-time **I/O** board and classify the resources according to the signal type through the adapter board and the signal conditioning carrier board; select the jumpers on the signal conditioning carrier board, centrally manage the ground wire, facilitate wiring and suppress interference; configure the channels to be conditioned through the signal conditioning carrier board and perform signal conditioning on the module; special signal simulation, such as linear oxygen sensor signal simulation, current-type wheel speed signal output conditioning, etc.

Fault Injection Unit

The fault injection unit can simulate common onboard electrical faults, such as open circuits and short circuits of wires, to test the fault diagnosis function of the electronic control unit. The fault injection unit generally adopts the distributed control mode. According to the pre-determined fault model, the upper computer controls the fault injection board through the fault injection operation terminal to simulate the corresponding onboard electrical faults.

Onboard Power Supply Simulator

The onboard power supply simulator uses EA programmable DC power supply PS8040-60 with a power capacity of 1.5 kW. The voltage and current of the DC power supply can be programmed, with a voltage programming range of 0–40 V and a current programming range of 0–60 A. Most of the hardware in the loop simulation

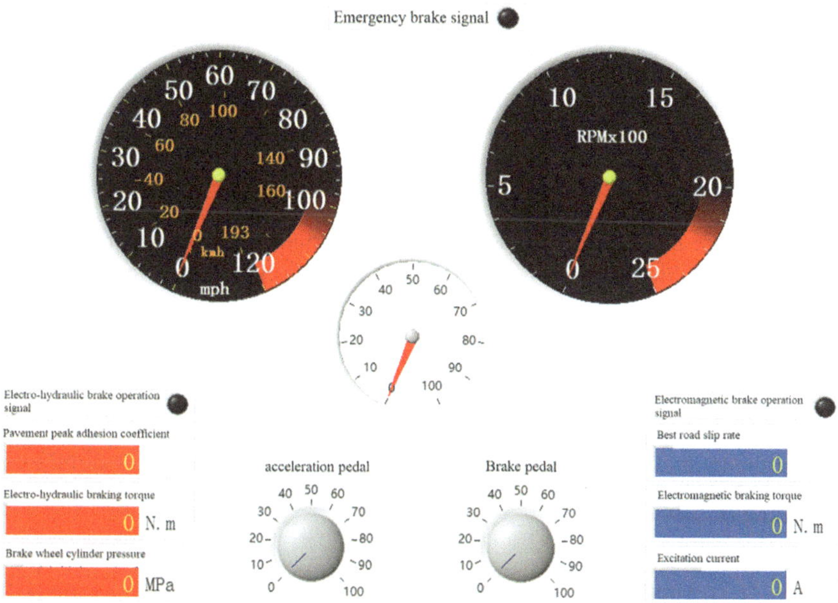


**Fig. 6.18** Test interface of the HILS platform for integrated brake system for electromagnetic and friction

platforms use voltage programming methods, and a 0–10 V analog signal controls the output voltage of 0–40 V.

#### 6.2.3.2 Test Management Engineering

The main function of test management engineering is to provide a human–computer interaction interface (virtual instrument and real-time interface), hardware management (hardware registration and upgrading), software model management (model download, link, data mapping), basic test management (test environment setting and test report generation). With the support of test management engineering, users can realize from the integrated operation of "model, hardware and application" greatly reduces the requirements for test operators, reduces the development workload of test engineers and improves test efficiency.

Test Application Software

As shown in Fig. 6.18, VeriStand can interconnect with the real-time operating system, establish virtual instruments by drag and drop, conduct dynamic data exchange with real-time programs, track real-time curves, adjust parameters online, real-time record data, playback real-time data, realize various functions such as display, modification and calibration of simulation model parameters, and build a monitoring and calibration system suitable for the test system. The test interface has obvious advantages.

Test Application Software

TestStand is used for the development, management and execution of test sequences. It provides a flexible and powerful framework for the development, management, and execution of test systems and is used to accelerate test and verification systems development. These test sequences integrate test modules written in any test programming language, and also standardize the execution process, report generation, database entry and connection with other enterprise systems. Test Stand can easily integrate independent test modules into the same test framework. Developers focus more on the development of test programs than the test framework required to build the test system.

#### 6.2.3.3 Test Application Software

The simulation model shall include not only the whole vehicle model, but also the necessary sensor model, actuator model and I/O interface model to realize the closed-loop operation between the simulation model and the hardware of the real-time simulator and the controller of the integrated brake system for electromagnetic and friction. Its principle is shown in Fig. 6.19.

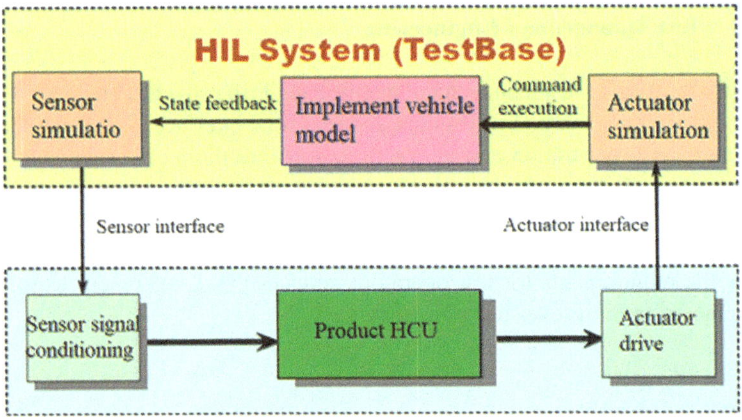


**Fig. 6.19** Schematic diagram of the HILS model's structure

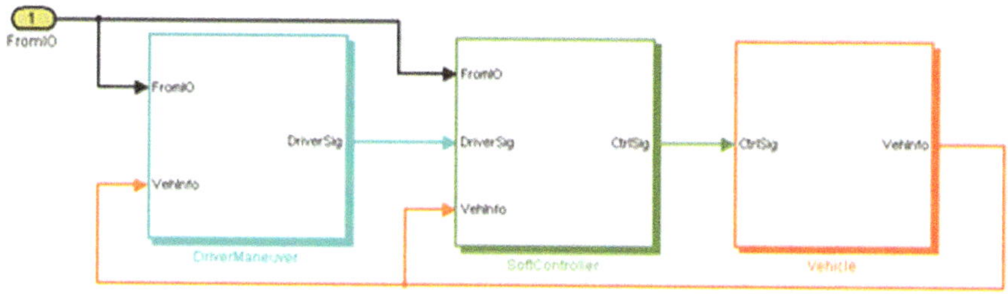


**Fig. 6.20** Schematic diagram of the vehicle model's structure

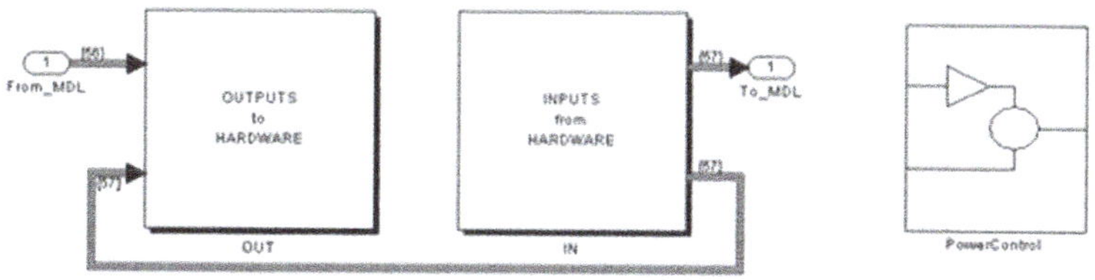


**Fig. 6.21** Schematic diagram of the I/O model's structure

## Vehicle Model

The whole vehicle model is mainly composed of three parts: Driver Maneuver module, Soft Controller module and Vehicle module, as shown in Fig. 6.20. Driver Maneuver module is used to simulate vehicle driving input, and the driving input sequence (including accelerator pedal, brake pedal, gear, target vehicle speed and other signals) needs to be set in the lookup table of the Driver Maneuver module in advance since the controller of integrated brake system for electromagnetic and friction is tested on the hardware in the loop simulation platform, other controllers on the whole vehicle need to be replaced by virtual electronic control units to realize the closed-loop control of relevant components. These virtual electronic control units are included in the Soft Controller module. The vehicle module consists of drive system, transmission system, integrated brake system for electromagnetic and friction, vehicle longitudinal dynamics and other modules are built according to the system structure diagram of the whole vehicle.

## I/O Model

The I/O model is mainly used for the mutual conversion between the physical values calculated by the simulation model and the electrical values of the hardware interface, as shown in Fig. 6.21. The I/O model includes OUT module, IN module and Power Control module. The out module converts the physical quantities, such as speed, displacement, switching state, into electrical signals and transmits them to the hardware system; the in module collects the electrical signals of the hardware system and transmits them to the vehicle model; Power Control module is used for remote control of onboard power simulator.

## 6.2.4 Antilock Braking Hardware-In-The-Loop Simulation

### 6.2.4.1 Anti-Lock Braking Model Test Verification

Figure 6.22 shows the adhesion coefficient and slip rate curves of several typical pavements. It is very difficult to directly simulate the adhesion coefficient and slip rate curves of various pavements on the rapid control prototype platform of integrated brake system for electromagnetic and friction; considering that the rapid control prototype platform can simulate similar to dry concrete pavement, the integrated brake system for electromagnetic and friction is carried out first. Then, based on the hardware in the loop simulation platform of integrated brake system for electromagnetic and friction built-in Sect. 7.3, investigate the effectiveness of the anti-lock braking coordinated control method proposed in this paper on dry asphalt icy and snowy pavements and docking pavements.

To compare the anti-lock braking performance of integrated brake system for electromagnetic, friction and electronic hydraulic brake, two braking modes are used for anti-lock braking control test respectively. Firstly, electronic hydraulic brake is used alone for anti-lock braking control, including estimating the peak adhesion coefficient and optimal slip rate of the road. As shown in Fig. 6.23, the simulation curve of the electronic hydraulic brake anti-lock braking control tends to be the same as the test curve, and the optimal road slip rate estimated during the simulation study and the test study is basically the same.

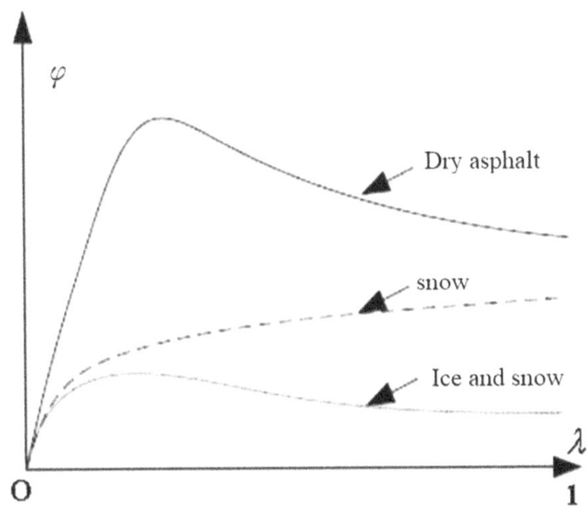


**Fig. 6.22** Curves of the adhesion coefficient and sliding rate in several typical roads

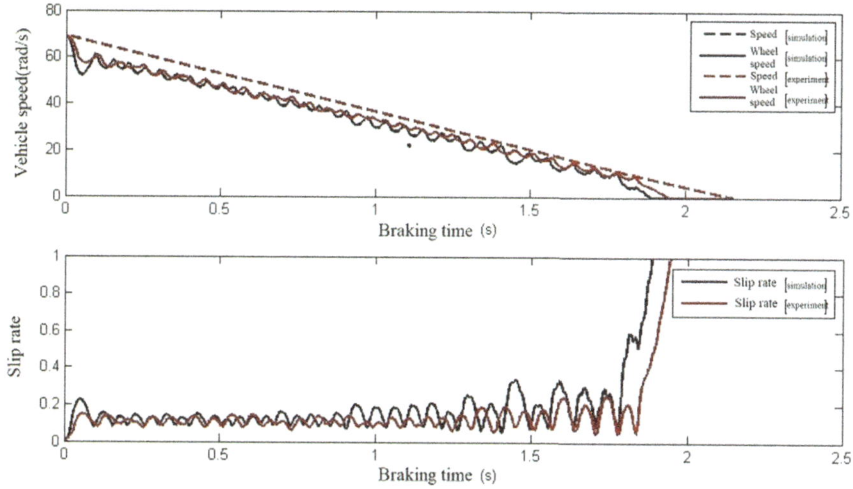


**Fig. 6.23** Curves of the anti-lock brake control simulation and experiment of the EHB

As shown in Fig. 6.24, on the dry concrete pavement with the pavement peak adhesion coefficient close to 1, the simulation curve of anti-lock braking control of integrated brake system for electromagnetic and friction is in good agreement with

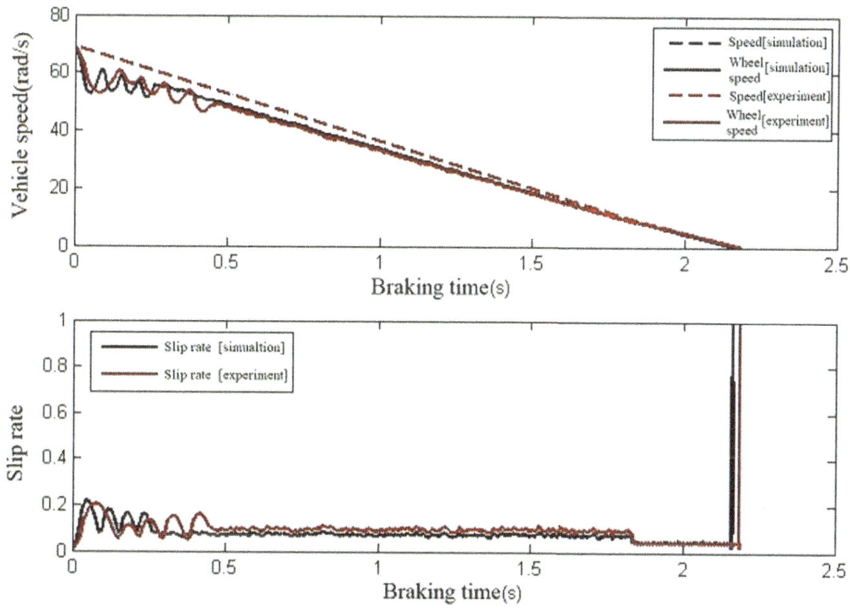


**Fig. 6.24** Curves of the anti-lock brake control simulation and experiment of the integrated brake system for electromagnetic and friction

the test curve. The second stage of anti-lock braking lasts until 0.44 s. There is a certain oscillation in the curve of wheel slip rate, which is due to the estimation of pavement peak adhesion coefficient by using electronic hydraulic brake alone. After that, it enters the third stage of anti-lock braking. The electromagnetic brake can accurately estimate the optimal slip rate and control the slip rate to remain at about 0.08. When the speed is lower than 10 km / h, turn off the electromagnetic brake and keep the output of electro-hydraulic braking torque unchanged. The simulation curve and test curve show that the actual slip rate of the wheel is reduced.

Combined with Figs. 6.23 and 6.24, it can be seen that on the dry concrete pavement simulated by the rapid control platform, the integrated brake system for electromagnetic and friction and electro-hydraulic brake can meet vehicle requirements anti-lock braking control. The time consumed in the anti-lock braking process of the whole vehicle is almost the same because the hydraulic regulator in the electro-hydraulic brake of the test object uses es of BOSCH company version 8.0 hydraulic regulator is replaced by ESP8.0. The high-speed solenoid valve in the ESP 8.0 hydraulic regulator has high hydraulic control accuracy.

#### 6.2.4.2 Hardware-In-The-Loop Simulation of Anti-Lock Braking Coordination Control

The research in Sect. 7.4.1 shows that the anti-lock braking model of integrated brake system for electromagnetic and friction established in the hardware in the loop simulation platform is accurate, and the next research work can be continued. In view of the requirements of investigating the impact of the performance of hydraulic control components in electro-hydraulic brake on the anti-lock braking of integrated brake system for electromagnetic and friction, for low-performance electro-hydraulic brake and high-performance electro-hydraulic brake respectively. Hardware in the loop simulation of anti-lock braking control for performance electro-hydraulic brake, integrated brake system for electromagnetic and friction.

As Fig. 6.25 shows, the control results of anti-lock braking on dry asphalt pavement are shown. The control accuracy of the inlet valve and outlet valve of high-performance electro-hydraulic brake is set to four times that of low-performance electro-hydraulic brake. The emergency braking effect of high-performance electro-hydraulic brake is significantly better than that of low-performance electro-hydraulic brake, and the emergency braking effect of low-performance electro-hydraulic brake is the worst. Although in friction and electric brake, low performance electro-hydraulic brake and electromagnetic brake are integrated in the magnetic integrated brake system, but the emergency braking effect of the integrated brake system for electromagnetic and friction is slightly better than that of the high-performance electro-hydraulic brake. This is due to the high accuracy of the optimal road slip rate estimated by the electromagnetic brake and the good effect of the electromagnetic brake to maintain the optimal slip rate. The use of the integrated brake system for electromagnetic and friction has high reliability. Using integrated brake system for electromagnetic and friction, high-performance electronic hydraulic brake and

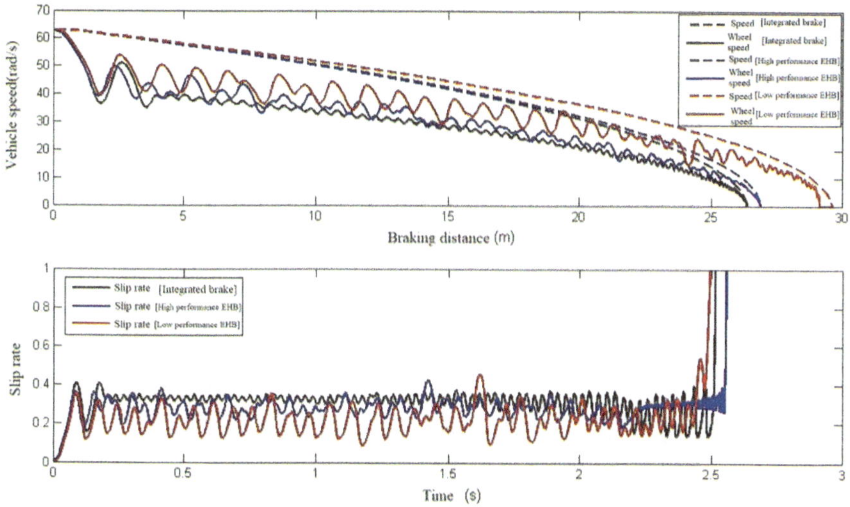


**Fig. 6.25** Curves of the anti-lock brake in dry asphalt pavement

low-performance electronic hydraulic brake, the best slip ratios are estimated to be 0.34, 0.32 and 0.28, respectively, while the initial setting value of the optimal slip rate of dry asphalt pavement in hardware in the loop simulation is 0.35. The ability to maintain the optimal slip rate can be quantified by the oscillation amplitude of the slip rate curve, the oscillation amplitude, the slip rate curve of friction, and the electromagnetic integrated brake system. Minimum, followed by high-performance electro-hydraulic brake. This shows that electromagnetic brake is significantly better than electro-hydraulic brake in control accuracy. Therefore, it has obvious advantages to use electromagnetic brake instead of electro-hydraulic brake in the third stage of anti-lock braking to maintain the optimal slip rate. The ability of the electronic hydraulic brake to maintain the best slip rate is related to the control accuracy of the high-speed solenoid valve. Due to the low performance of the electronic hydraulic brake, the control accuracy of the inlet valve and the outlet valve is relatively low. so the oscillation amplitude of the variation curve of slip rate with time is also the largest.

Figure 6.26 shows the results of anti-lock braking control on icy and snowy roads. Different from dry asphalt roads, the braking distance of high-performance electronic hydraulic brake is slightly lower than that of the integrated brake system for electromagnetic and friction. The braking intensity required for emergency braking on icy and snowy roads is low, resulting in the reduction of the electro-hydraulic braking torque, and the low-performance electro-hydraulic brake has a poor ability to maintain the target braking torque. Therefore, it can be found that the use of friction and electromagnetic When the integrated brake system performs anti-lock

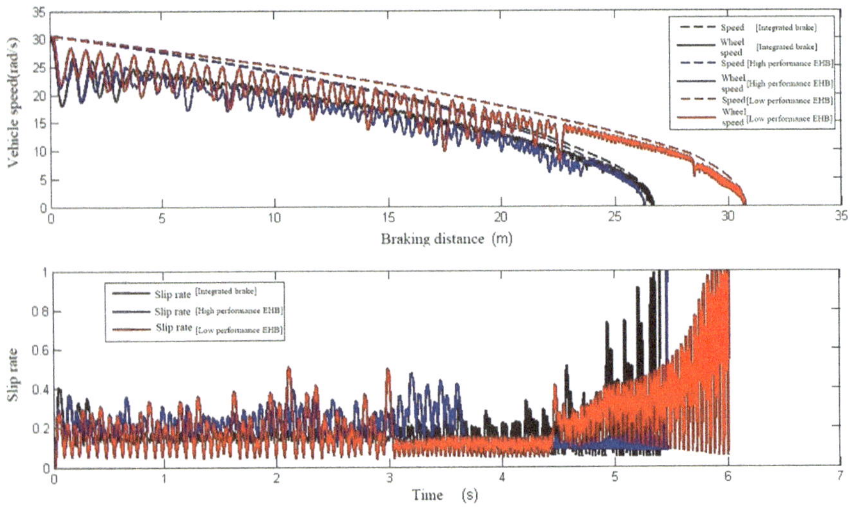


**Fig. 6.26** Curves of the anti-lock brake in ice and snow pavement

braking control, when the wheel speed is lower than 10 km/h, the electromagnetic brake is removed, and the oscillation amplitude of the slip rate curve is significantly increased. By comparing Figs. 6.25 and 6.26, it is shown that high-performance electronic hydraulic brake and low-performance electronic hydraulic brake on dry asphalt pavement slip rate versus time curve oscillates better than icy and snowy roads. This is due to the emergency on low-adhesion roads. The braking intensity required for braking is lower than that of a high-adhesion road. Under the same high-speed solenoid valve control accuracy, the ability of electronic hydraulic brake to estimate the optimal slip rate of the road surface and maintain the optimal slip rate becomes worse.

The docking pavement is to verify the anti-lock braking control performance of high-performance electro-hydraulic brake and integrated brake system for electromagnetic and friction when jumping from dry asphalt pavement to ice and snow pavement. As Fig. 6.27 shows, the anti-lock braking control results on the docking pavement show that the emergency braking effect of high-performance electro-hydraulic brake is worse than that of integrated brake system for electromagnetic and friction, which is still estimated for both the difference between the optimal slip rate and the ability to maintain the optimal slip rate.

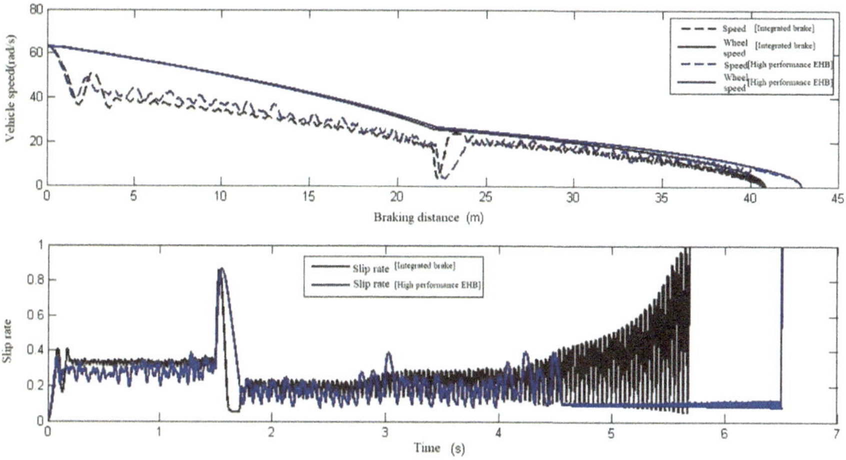


**Fig. 6.27** Curves of the anti-lock brake in docking pavement

# References


1. David R, Alla H (2001) On hybrid Petri nets. Discrete Event Dyn Syst 11(1–2):9–40
2. Alla H, David R (1998) Continuous and hybrid Petri nets. J Circuits Syst Comput 8(01):159–188
3. Maofei Z, Wuwei C, Guang X (2011) Decoupling control of automobile chassis based on neural network inverse system method. Trans Chin Soc Agric Mach 12:13–17
4. Wu J, Wang X, Li L et al (2018) Hierarchical control strategy with battery aging consideration for hybrid electric vehicle regenerative braking control. Energy 145(145):301–312
5. Chen J, Yu J, Zhang K et al (2018) Control of regenerative braking systems for four-wheel-independently-actuated electric vehicles. Mechatronics 50(50):394–401
6. Peng J, He H, Xiong R (2016) Rule based energy management strategy for a series–parallel plug-in hybrid electric bus optimized by dynamic programming. Appl Energy 185(pt.2):1633–1643
7. Hailong S (2014) Development of pure electric vehicle hardware-in-the-loop test system based on virtual test technology. Shanghai Jiaotong University

The manufacturer's authorised representative in the EU is Springer Nature Customer Service Centre GmbH, Europaplatz 3, 69115 Heidelberg, Germany. If you have any concerns regarding our products, please contact ProductSafety@springernature.com

Printed and bound by CPI Group (UK) Ltd, Croydon, CR0 4YY
30/09/2026
02225307-0010